Modern Methods of Systems Engineering

With an Introduction to Pattern and Model Based Methods

Joe Jenney with:

Mike Gangl, Rick Kwolek, David Melton, Nancy Ridenour, and Martin Coe

Copyright © 2010 Joe Jenney

All rights reserved.

ISBN-13: 978-1463777357

DEDICATION

To all the systems engineers the authors have had the pleasure of working with and learning from over many years.

CONTENTS

	Acknowledgments	vi
	Preface	1
1	Introduction and Overview	3
2	Engineering Methods for Product Development	16
3	Introduction to Systems Engineering	27
4	Processes and Tools for Planning a Program	38
5	Introduction to Pattern Based Systems Engineering	67
6	Processes and Tools for Defining a System	76
7	Quality Function Deployment (QFD) In System Engineering	133
8	Selecting the Preferred Design	160
9	Processes and Tools for Verifying Technical Performance	179
10	Processes and Tools for Risk Management	187
11	Introduction to Model Based Systems Engineering	197
12	Integrating Modern Methods for Faster Systems Engineering	211
	References	221
	Acronyms	224
	Index	227
	About the Authors	234

ACKNOWLEDGMENTS

In my 40 year career in the defense and aerospace industry I had the good fortune to know and to work with some outstanding systems engineers. Anything I know about systems engineering I learned from others far more expert than me. In the 1970s I worked in the Strategic Technology Office of the Defense Advanced Research Projects Agency (DARPA). It was there that I learned that experienced systems engineers can not only perform the basic systems engineering processes associated with system development but can apply these techniques to mission analysis to ensure that the right technologies for the right systems solutions are developed for future needs. DARPA gets a lot of credit for its successes over the years and I learned that experienced systems engineers in the broad defense and aerospace community help keep DARPA managers on the right track.

In my career in industry I worked with systems engineers involved with numerous complex system developments. I learned that effective systems engineering is essential to the success of any development program. I also learned that although systems engineering processes are not always obvious to the inexperienced these processes can be learned and applied if engineers have basic knowledge of engineering and of the technologies and existing systems in a mission area.

In this book my coauthors and I explain the fundamentals of systems engineering processes and introduce several methodologies that have emerged in the past decade that enable systems engineering to be accomplished faster and with higher quality. The material presented here grew out of a training program that I developed with the help of many experienced systems engineers in the 1990s. We started with the practices and tools used by those we judged to be the best systems engineers we knew in our company and in our customer's organizations. We then added selected standard methodologies taught in popular handbooks of systems engineering. Finally we have introduced new and emerging methods that greatly reduce the time and effort needed for effective systems engineering. The methodologies and tools taught here have been used extensively and proven to be effective by many experienced systems engineers.

Specifically I would like to acknowledge the contributions of my coauthors. Mike Gangl contributed significantly to the training program the book is based on and participated with me in training many times. In addition to Mike's teaching me much of this material he wrote the section on Systems Engineering for Defense Acquisition and reviewed other chapters.

I have had the pleasure of working with Rick Kwolek on front end systems engineering for complex systems on a very fast paced schedule. He is dedicated to developing complete and accurate system requirements and has developed some of the pattern based systems engineering ideas presented in this book. Rick wrote much of chapter six.

Every successful engineering organization has a few "go to" experts for difficult problems. David Melton was the "go to" expert for mechanical, thermal and complex system problems for many years in the systems engineering organization I managed. He is also an accomplished executive and an expert in the application of Quality Function Deployment and Taguchi Design of Experiment methods. David wrote the chapter on using Quality Function Deployment for systems engineering.

When I was searching for an author for the chapter introducing model based systems engineering everyone I queried recommended Nancy Ridenour. I was delighted when Nancy agreed to write this chapter in parallel with her being the manager of software on a large complex program.

Almost all of my experience and the experience of the coauthors cited above has been in aerospace and defense businesses. We are aware of the challenges of applying effective systems engineering in the development of commercial products but have little direct experience. We were fortunate to find that Martin Coe, who has extensive experience in both developing commercial products and in being an advocate for systems engineering for commercial products, was willing to contribute the section in chapter one on Systems Engineering for Commercial Product Development.

PREFACE

Systems engineers face new constraints in the 21st century. Customers want ever faster and cheaper system development. System complexity continues to increase due to customer demands and market opportunities. The increased complexity is facilitated by the continuing improvements in the technologies of materials, sensors and digital electronics. Skilled engineers and engineering managers are retiring faster than replacements are ready due to the management practices of flattening and leaning organizations in the last decades of the 20th century. Engineering teams are spread across multiple sites.

System engineers have new tools and processes to address the new constraints. The internet and intranets enable effective and rapid communications across multiple sites. Computers and internet communication tools are relatively cheap. Powerful computer aided design and system simulation tools are available. Integrated design and design documentation processes are maturing. Some integrated design and manufacturing tools are available and more are likely to become available. The potential for reliable end to end document management is at hand. Model and pattern based system engineering methods are emerging. The central question is how systems engineers and their organizations use these new methods and tools to relieve the constraints of this new century.

This book reviews the fundamentals of systems engineering and attempts to show how modern decision management and model and pattern/template based methods can be integrated and applied to the fundamentals to achieve reuse of systems engineering as well as reuse of hardware and software designs. This can both shorten the time required for the systems engineering work and increase the quality of the systems engineering documentation necessary for design and development of modern complex systems. A second objective is to show how modern information technology tools used by integrated product development teams can enable the return of the role of chief designers and thereby reduce information latency. This further reduces the time and cost of systems engineering work.

It is the author's claim that executing proven fundamental systems engineering processes with an approach based on the integration of decision management, model based systems engineering, pattern/template based systems engineering, modern information technology tools and carried out by a chief designer leading integrated product development teams results in minimizing the cost and schedule of systems engineering. This combination is the best approach we know for addressing the demands of 21st century systems engineering.

A third objective of this book is to present material in a way that both facilitates learning by the readers and enables the readers to begin the development of pattern/template based methods for their organizations by faithfully doing the exercises in the book. This book in intended to be studied over a period of time and time should be devoted to complete the exercises; it is not intended to be a quick weekend read. The material is structured to be a complement to other systems engineering handbooks, especially those that are available freely on the internet like the DoD *Systems Engineering Fundamentals* handbook and the NASA *Systems Engineering Handbook*. The material also complements the ICOSE *Systems Engineering Handbook* and systems engineering processes recommended by IEEE.

The authors believe that systems engineers should study multiple sources and multiple methods and then adopt and adapt methods that work for their organizations and their systems development efforts. No one set of tools or methods applies equally well for all types of products and systems or is likely to be found best by all systems engineers. The material in this book is deliberately organized differently than the material in other handbooks. The intent is that readers who study this material and the same material presented differently in other handbooks will be stimulated to think more deeply about the information presented. It is hoped that this leads readers to develop their own refinements of the methods presented and perhaps even develop the better methods that will be needed to effectively develop the even more complex systems expected in the decades ahead.

Little mention is made of the many commercially available software tools necessary and useful for modern systems engineering. These tools are evolving rapidly and it would be difficult to provide a fair appraisal of many of them. Tools are often best suited to systems of a certain size or type so it isn't possible to recommend tools that are best for every system development. Where a tool is mentioned it is only as an example and should not be viewed as an endorsement of the tool over competing tools. Readers that learn the fundamentals presented here and in the available handbooks cited above will be prepared to choose and learn how to use tools that apply well to their system developments.

Systems engineering includes the systems engineering process along with the time phasing of systems development, life cycle integration and other management processes, e.g. as defined by the DoD *Systems Engineering Fundamentals* handbook. This book primarily addresses the systems engineering process because the intent is to introduce new process methods that reduce the cost and improve the quality of the process and because most organizations prefer to tailor management processes to their particular needs.

1 INTRODUCTION AND OVERVIEW

1.0 Objectives of This Training Material

Systems engineering is an ever evolving engineering discipline. In the late 1990s new methods emerged that enable shorter product development cycles while at the same time improving the quality of systems engineering documentation that is critical to the development of high quality products. Systems engineers and managers of product development have an excellent selection of handbooks and books available from US Government agencies and professional organizations that treat systems engineering fundamentals. Unfortunately the most useful ones have not incorporated the newest methodologies. The intent of this work is to provide an introduction and training guide in these new methodologies so that product development organizations can realize significant reductions in the systems engineering portions of product development and achieve higher quality products. Most of the book is concerned with the fundamentals of systems engineering because understanding the new methods isn't possible without understanding the fundamentals. The specific objectives of this material are to:

- Teach modern systems engineering methods that support product development on short schedules and at low development costs
- Teach methods and tools in formats compatible with standard systems engineering references to enable students to continue to study on their own after finishing this material
- Provide initial steps toward systems engineering templates for rapid product development

This material will not:

- Teach the domain knowledge specific to any particular system
- Substitute for the years of experience and engineering education needed by systems engineers
- Teach the detailed systems engineering processes and project management topics covered very well by the complementary books cited above
- Discuss modern methods of managing the entire product life cycle, i.e. Product Life Cycle Management (PLM) other than the systems engineering portions.

The material presented is not for systems engineers alone. It should be studied by managers responsible for systems engineering including functional engineering managers and product development managers. It does little good to have systems engineers trained in the most modern and efficient methods if their managers do not understand the value of modern methods and do not support the implementation and use of these methods. It is also valuable for other engineers involved in product development to learn these methods because many apply to design engineering as well as systems engineering. Note, here the term product includes systems, products and complex services; all of which benefit from effective systems engineering.

1.1 Sources of the Training Material

The material presented here has evolved over a number of years and from a number of sources in addition to those cited in the references. The most important sources include:

- "Best Practices" of engineers on successful projects in the authors' experience
- The Department of Defense Systems Management College publication *Systems Engineering Fundamentals*[1-1]
- NASA *Systems Engineering Handbook* Sp-2007-6105,Rev 1[1-2]
- IEEE Std 1220-1998 *Standard for Application and Management of the Systems Engineering Process* [1-3]
- *INCOSE* (International Council on Systems Engineering) *Systems Engineering Handbook* [1-4]
- GPR 7120.5A *Systems Engineering* (GPR is NASA Goddard Procedural Requirements) [1-5]
- Technical papers published by the Vitech Corporation [1-6]
- The Systematica™ Methodology of ICTT System Sciences [1-7]
- Studies of *"Integrated Concurrent Engineering"* available on the internet at various sites; for example *"Observation, Theory, and Simulation of Integrated Concurrent Engineering: Grounded Theoretical Factors that Enable Radical Project Acceleration"* by John Chachere, John Kunz, and Raymond Levitt [1-8].

The intention of the authors is that this material complements the DoD's *Systems Engineering Fundamentals*, the NASA *Systems Engineering Handbook* and IEEE 1220 so that students can continue to study using these sources after completing this material. However, the referenced sources are not a substitute for studying this material because concepts critical to short cycle time and low cost systems engineering are presented here that are not in the DoD, IEEE and NASA references. Thus this material plus DoD's *Systems Engineering Fundamentals*, the NASA SE Handbook and IEEE 1220, or this material plus the INCOSE *Systems Engineering Handbook* can be considered a handbook for top level systems engineering. Readers may wonder why the international standard ISO/IEC 15288 *Systems and software engineering —System life cycle processes* [1-9] is not referenced. The reason is that this standard is more like a policy and procedures document for all the management and technical processes relating to product development than a handbook for systems engineers.

A cautionary note on systems engineering nomenclature is necessary for readers. Systems engineering nomenclature is not standardized although there are efforts underway. Therefore it's not possible to exclusively use standard nomenclature. Nomenclature consistent with the INCOSE *Systems Engineering Handbook* and DoD's *Systems Engineering Fundamentals* is used where possible. Inconsistency in nomenclature isn't usually a problem but some may find it confusing. the authors apologize if unfamiliar nomenclature is used or the nomenclature deviates from the documents cited. Also, in some cases nontraditional nomenclature is deliberately used to force readers to think about system engineering in new or in more fundamental terms.

1.2 Approach to Introducing New Methods

In principle this book could be divided into two parts in which the first part treats the systems engineering fundamentals necessary to understand and implement modern methods and the second part treats the modern methods. However, this leads to delaying the introduction of concepts that are important for using the traditional fundamental methods effectively. Therefore the approach chosen is to cover the fundamentals first but introduce modern methods wherever it is useful to employ the modern methods in implementing fundamental processes.

Thus, with the above caveat, the approach that is followed is to:

- Summarize how engineering methods have evolved and led to modern systems engineering
- Review the fundamentals necessary to understand modern methods
- Describe methods and tools in formats compatible with standard references
- Provide exercises for the students that help review the material and become initial steps toward reusable templates specific to their work

Also this book seeks to achieve three understandings in addition to introducing Pattern and Model Based Systems Engineering; these are:

- How and why product development has evolved
- The fundamental systems engineering activities that are critical to the success of product development
- The roles and responsibilities of systems engineers and other team members in modern product development

In addition to the three understanding listed, the book seeks to explain important processes and tools available to engineers for reducing product development time, show some practical examples of the processes and tools and provide the reader exercises to practice applying these processes and tools. Before beginning to address these three understandings there is some introductory material that needs to be explained.

Understanding how and why product development has evolved over the past century is helpful in understanding modern methods. It also aids in encouraging the use of modern methods. Understanding the roles and responsibilities of systems engineers and other team members helps promote working as part of a team and encourages system engineers and their managers to involve design engineers, suppliers and manufacturing people early in product development. Understanding system engineering activities encourages focusing on the product under development, not the development process and helps understand the implications of the complexity of modern systems and how to deal with this complexity.

1.3 Recommended Study Approaches

The material is structured similar to a text book but with some important differences. It can be studied by individuals but it is more effective to study as a team of four to six people. If there are more than six people studying then break up into more teams. The best and most effective study approach is for a team that is just starting a new product development project to study this material together and to use the recommended exercises to develop required documentation for their project. Thus this material becomes an adjunct guide during the systems engineering phase of their work. This approach requires a facilitator so one individual should learn the material first and act as facilitator for the development team later.

If only one individual or one team is studying then save the results of the exercises electronically so that these results can be reused on future product developments. If more than one team is studying it helps to do the exercises either on flip charts to facilitate sharing each team's results or in a facility that enables electronic sharing of exercise results.

If a team is beyond the start of a new project the second best study approach is for the team to invent a straw man product and work through the material developing documentation for the straw man product. It is critical that the straw man product be as close as possible to a typical product the organization develops. The reason is that the documentation developed during the study of this material is intended to become the first generation of templates that are reusable on future product developments.

If an individual is studying this material on their own then either the straw man product approach can be used or choose a previous product from the organization's development history for the exercises.

1.4 Rationale for Using Modern Systems Engineering Methods

The objective of product development is to achieve the shortest practical time from concept to production, i.e. the development cycle time, at the lowest cost consistent with the required product quality. Numerous books and papers have discussed the importance of being first to market. Or if not first, getting to market with cost or feature advantages. Unfortunately, product development managers too often attempt to shorten the development cycle by limiting the systems engineering effort or terminating the systems engineering too early.

Study after study has shown that cutting corners in systems engineering to save a small fraction of development cost or time results in problems in the later stages of the development cycle, like system test and initial production, which increase development costs many times greater than the costs saved earlier. The reason this happens is also revealed in such studies of costs over the development cycle. The systems engineering phase of product development typically costs about 10% of the total product development. However, it is found that often 80 to 90 % of the system costs are determined by decisions made during the first 10% of the development cost. There are two important consequences of this finding.

One is that it is possible to explore many design approaches during the systems engineering phase at very low cost. Only a portion of the systems engineering costs are spent exploring design approaches; the rest is spent on requirements analysis, documentation and communication. Thus it's possible to explore perhaps twice as many system design approaches for an increase in total development cost of only about 2%. Since doubling the number of design approaches explored significantly increases the likelihood that a higher quality or lower cost design approach is found it is well worth the extra expense.

The second important consequence is that systems engineering occurs at the beginning of the development cycle and no product development manager wants to deviate from a planned budget in the earliest stage of a project. It takes exceptional courage for a project manager to overrun the planned budget during the systems engineering phase no matter how well the manager understands the likely consequences. It is just human nature to hope that past lessons learned won't apply to the current project. Thus managers are tempted to cut corners during systems engineering rather than conduct a thorough job even though doing a thorough job is highly likely to save costs in the long run.

There is a way out of the manager's dilemma. The way out is to use modern systems engineering methods that dramatically shorten the time and cost of the systems engineering phase of product development and improve the quality of the systems engineering work at the same time. This sounds too good to be true but it has been proven to work.

If the modern methods are so effective why isn't everyone using them? Good question. One reason is that organizations are reluctant to try new methods and another is that these new methods can be costly to implement. However, there is a very cost effective way to implement these new methods. But before we begin to explore new methods it's important to understand how to best use traditional methods. The experience and discipline of using traditional methods properly carries over to helping achieve the best use of more modern methods.

1.5 Keys to Effective Product Development

Effective product development can be defined as achieving the shortest possible development cycle time at the lowest possible cost consistent with the desired product quality. Some of the most important elements of effective product development are:

1. Effective communications between managers, systems engineers, design engineers (here design engineers includes all the specialty engineers; electrical, mechanical, software, quality, etc.), suppliers and

personnel responsible for production as indicated in Figure 1-1. If customers are paying for all or a portion of the development then effective communication is necessary between all stakeholders.

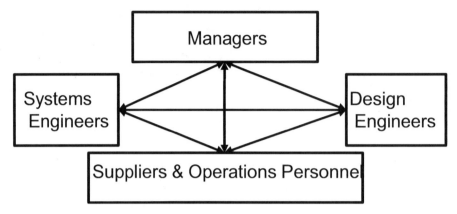

Figure 1-1 Effective communications must be maintained between all personnel involved in product development during all phases of the work.

2. Never having to redo something that was done correctly on a previous development program and satisfies the requirements of the current program.

3. Using modern methods that have proven effective in minimizing information latency. (Information latency is the time between when information is available and when it is communicated to those that need the information to do their work.)

4. Working as a team and using good teamwork practices.

5. Developing schedules and assigning resources that accommodate working as a team.

6. Using tools and methods that facilitate preventing problems rather than finding and fixing problems.

7. Maintaining the discipline of always conducting peer reviews of engineering work and providing time for peer reviews in project schedules.

8. Giving task leaders and engineers the authority to seek input from anyone.

9. Technical, financial and market value statements must be considered at each stage of the development.

Elements 1, 7 and 8 are tied to leadership and can be studied in depth in the book *The Manager's Guide for Effective Leadership*[1-10]. The five elements between 1 and 7 relate to methods and tools; of course methods and tools can contribute to effective communications as well. The first six important elements are only achievable if systems engineering methods and tools are used in product development in addition to good design practices and tools. Element 9 is a statement of the basic discipline teams must follow to keep product development on track.

Enterprises doing work for the U. S. Department of Defense and NASA typically use systems engineering methods and the systems engineers in these enterprises need to understand how systems engineering is used in Government procurement agencies. Disciplined systems engineering is not always followed in enterprises engaged in commercial product development. The reasons why systems engineering methods should be used in commercial product development and the methods used by Defense agencies involved in procurement activities are addressed next.

1.6 Systems Engineering for Commercial Product Development[1-11] **1.6.1 Introduction-** Why is systems engineering required for commercial product development? This question is answered by illustrating the correlation between commercial product development and the application of systems engineering. More specifically this analysis:

- Defines the main business objective of commercial product development
- Introduces a product profit model that illustrates the relationship between this objective and 5 characteristics critical to successful commercial product development
- Utilizes the model to illustrate how product profit can be adversely affected if these 5 critical characteristics are not optimized
- Identifies the relationship between systems engineering and these 5 characteristics critical to successful product development

1.6.2 Commercial Product Development's Main Business Objective - Regardless of the commercial product being developed or the industry it will be sold, installed or operated in all commercial product development has a common business objective. That being <u>to ensure each product development project earns a sufficient profit</u>. Whereas the definition of sufficient may vary among companies, products and projects the constant is that no company can continue to perform product development for any length of time without earning profits from the products they develop.

Figure 1-2 is an example of a model that illustrates possible product revenue generated and lifecycle costs incurred over the lifecycle of a product development project. Since

Profit = Product revenues – Lifecycle costs

such a model can be used to track project profits and therefore progress toward meeting a business objective.

Development of the actual model begins early in a project when the lifecycle costs and size/ shape of the product revenue area are estimated yielding an estimated profit; use of the model occurs during the project when these parameters are measured yielding a profit snapshot; and after product end of life (EOL) when these parameters can be confidently determined to yield a final profit value.

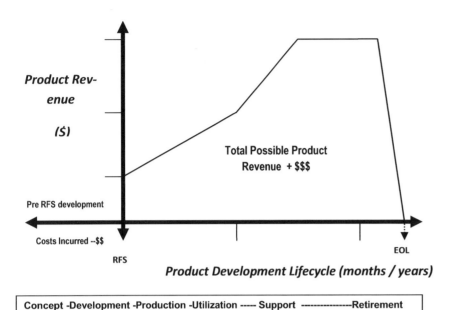

Figure 1-2 The Product Profit Model tracks life cycle costs and product revenues over the life cycle of a product.

In this model the *area under the curve* represents the *Total possible product revenue (+$)* generated from any product and *lifecycle costs (-$)* both measured throughout the product development lifecycle. From the model note:

- **At Release for Sale (RFS)** the developed product is released for sale and product revenue generation begins
- **At End of Life (EOL)** product obsolescence/retirement is imminent and product revenue generation ends
- **Lifecycle Costs** are being incurred throughout the product development lifecycle (from the beginning of the development to the end of life)

1.6.3 How is the Profit Objective Met? - As the model illustrates meeting the profit objective means ensuring that the total product revenue is maximum and that lifecycle costs incurred over the product development lifecycle are minimum. As the number of factors affecting product revenue and lifecycle cost during product development are too numerous to manage these are consolidated to five critical characteristics for this discussion. Optimization of these five characteristics is imperative to maximize product revenue and minimize lifecycle costs for any commercial product development project (tradeoffs between characteristics are needed to optimize the highest priority characteristics for each specific project).

A brief introduction of these five characteristics within the context of product development begins to reveal how the size and shape of the product revenue curve and lifecycle cost value can be adversely affected for any project.

Meeting stakeholder needs - Every effort must be made to completely and accurately identify, define and satisfy project stakeholder needs as reflected in a comprehensive set of product requirements. Inadequate, missing or poorly defined requirements can affect customer satisfaction (final slope) or market size (plateau).

Minimizing Development Time to Market (Schedule) - The time needed to develop a product from concept to release for sale (RFS). Generally, increasing the time to market increases the lifecycle costs and reduces the revenue generation period (initial slope and less curve width).

Maximizing Product Performance - Product performance, measured from the customer/user perspective, covers many design characteristics such as reliability and quality. Poor product performance can affect the size of customer base (plateau) or product longevity (final slope).

Minimizing Lifecycle Costs - The total product cost to the developer throughout the product development lifecycle. Excessive lifecycle costs can affect the product market price (initial slope) and/or size of customer base (plateau) and directly impact profit margin.

Optimizing Risk - Product development risks can be cost, schedule, performance, requirements or programmatic related in nature. Performing effective risk management throughout product development is an integral part of optimizing these critical characteristics. Nonexistent or ineffective risk management can affect any or all aspects of the curve.

The fate of many product profit margins is doomed early in the project (often long before RFS) due to the lack of attention to the five critical characteristics during product development. The model described here reveals that not paying attention to these characteristics can easily result in products that do not realize a profit margin and therefore do not meet business objectives.

1.6.4 Systems Engineering is Directly Related to The Critical Five Characteristics - Any optimization of the five characteristics involves the application of systems engineering methods to product development projects. If any one of these characteristics is measured, controlled, or managed during product development or special tasks/functions relating to these areas are performed then systems engineering methods are being applied to product development. Table 1-1 clarifies how systems engineering is related to these critical characteristics with a brief list of systems engineering functions. This is why systems engineering is required for successful development of commercial products.

Critical Characteristic	*Related Systems Engineering Function*
Meeting stakeholder needs	Voice of the customer, stakeholder analysis, requirements engineering, design verification, design validation
Minimizing schedule	Resource management, project schedule development, project scoping, work breakdown structures, earned value analysis, quantitative analysis
Maximizing product performance	Design for reliability/quality, requirements engineering, system integration, system architecting, software engineering, system analysis
Minimizing lifecycle costs	Lifecycle cost analysis, earned value analysis, business case analysis, engineering accounting
Optimizing risk	Risk management, root cause analysis, process validation, team development, six sigma processes

Table 1-1 Critical business characteristics are related to systems engineering functions.

1.7 Systems Engineering for Defense Acquisition

Whereas systems engineering is needed to develop any new system or improve on existing products, there are unique activities and nomenclature for those engineers who are supporting US Department of Defense (DoD) acquisition programs. DoD has recognized that the systems they procure are much too complicated to not address quality systems engineering practices. Additionally, if these procured systems do not perform as required or fail in operation, then the consequences may be severe including failed US policies or even injury or death to our military men and women. Therefore, it is important that engineers and program managers who are working on government projects understand the system engineering processes defined by the DoD and how these relate to their own product development systems engineering processes.

The standard by which the DoD acquires new systems, from planes to ships to radios and beyond, is Department of Defense Instruction (DoDI) 5000.02, *Operation of the Defense Acquisition System*[1-12]. Figure 1 of the "5000" process shows the primary stages of the acquisition cycle, which include pre-system acquisition, system acquisition and sustainment. The DoD acquisition community has statutory and regulatory requirements to follow this process to procure new military systems. But there are two important exceptions.

The first exception is for research and development. Government organizations such as the Defense Advanced Research Projects Agency (DARPA) and the military laboratories issue research and technology development contracts to other government agencies and industry contractors. These contracts do not follow the 5000 process. Typically, the systems engineers receive a Statement of Objectives (SOO) and perhaps a set of target performance requirements. In some cases, such as DARPA with their go-no go criteria, further funds are predicated on achieved performance. This is good since no program manager or systems engineer should sign up to meet hard, budget linked requirements for technologies that are not yet demonstrated. Of course, companies are frequently tempted to do so to win the contract. So it is important that the engineers, even on these technology development projects, follow some form of a systems engineering process and ensure all stakeholders understand risks and mitigation tasks. A company's profit and reputation depend on it.

The second exception is a more recently popular acquisition approach called Quick Reaction Capability, or QRC. For these types of contracts, the system developer is provided with a set of requirements the military need to satisfy to field an operational capability. A QRC is typically justified to either demonstrate the utility of a new technology or mission, or to satisfy an immediate need of the warfighter. This is especially important in times of war. The largest, most recent example is fielding technologies to defeat improvised explosive devices (IEDs). With a QRC program, the government elects to forgo the 5000 process and have the contractor build and field the equipment on a best effort basis.

Of course, the degree to which the requirements are met still has a significant impact on whether the customer is satisfied. Here, it is extremely important that the engineers maintain some level of systems engineering discipline in a highly accelerated schedule. Managing performance and expectations are even more critical when working "under the gun." A big issue for many DoD acquisition programs in today's environment is having the Pentagon leverage a 5000 process while imposing a QRC schedule. Everyone wants the latest and best capability; and in time of war, this is amplified. These competing requirements, full acquisition process vs. schedule, are not easily solved and require continued systems improvements to the acquisition process.

Back on the subject of the 5000 process; it is important to point out some of the key systems engineering products the government is required to produce and how they relate to engineers and their programs. The whole DoDI 5000.02 process is not covered here since that is a separate, extensive topic. Training can be found through the Defense Acquisition University (www.dau.mil). Typically, this training is only necessary for government systems engineers. There are three important items for engineers who work on government contracts to develop systems.

The first is the flow of requirements. It is important that a contractor systems engineer understand the way the program requirements have flowed down through the government process. For a typical military acquisition program, the primary requirements are captured in a Capability Development Document (CDD). The CDD contains the government agency's Key Performance Parameters (KPPs), Key System Attributes (KSAs) and any additional attributes they believe are necessary to have the system satisfy their objectives. The CDD is usually worded in language from the user community rather than from technical personnel. The government organization that is responsible for the acquisition then must procure a system that satisfies these KPPs and KSAs. The systems engineers of the government acquisition organization must translate these CDD requirements into technical and operational requirements. These are documented in a System Requirements Document (SRD). Many government systems engineers now use software tools such as DOORS™ to track the linkage of the CDD to the SRD. Typically, the government may release either a top-level SRD or a complete SRD. The difference is the amount of engineering maturation that occurred to derive the SRD requirements. If the agency doesn't have all the expertise and time at their disposal they may elect to go with a top-level SRD and work with the contractor to mature the requirements. Prior to entering the Engineering and Manufacturing Development phase of the program, the

SRD will be worked into a final System Specification (sometimes called a Weapon Systems Spec, or WSS). The WSS is intended to resolve all To Be Determined (TBDs), To Be Reviewed (TBRs), and no longer contain Threshold (T) and Objective (O) requirements. It is expected that the WSS contains the specific requirements that the government and contractor will comply with during fabrication and test.

The second important DoDI 5000.02 document to understand is the Systems Engineering Plan (SEP). Like Systems Engineering Management Plans (SEMPs) that many contractor companies create, the SEP describes the engineering processes, tools, plans and schedules as determined by the government acquisition engineers. The SEP may be reviewed as high as the Pentagon, depending on the program's acquisition category. It is considered a living document that changes as the program matures. It takes on different information as the program progresses through development maturity. Although the government continues to own and maintain the SEP they expect the contractor's program management and engineers to not only contribute to the maintenance of the document but to synchronize any of their own SE processes and plans with it as well. It should be noted that the SEP must be updated and provided at key milestones throughout the acquisition process.

The third important item to reference is the DoD Architecture Framework (DoDAF). This is a series of pictorial and graphical representations of the system being developed. It can be thought of as the documentation of the functional and physical architectures. The top level diagram of the DoDAF is the Operational View -1 (OV-1). This is typically a concept type picture that shows how the system may be operated. Essential elements are shown in the OV-1 along with external elements that the system is expected to interface with. There are additional Operational View documents to further describe these system and external elements and how they are intended to operate.

Additionally, the DoDAF has other diagrams and documents under the major categories of Systems and Services View (SV-x) and Technical Standards View (TV-x). The SV articles should be closely watched since it typically contains specifics on interfaces to external entities and is used to derive Interface Control Documents. Note: there is also a higher level All View (AV) but systems engineers may not find those as beneficial.

The DoDAF is started by the government operational agency to define the system and describe its intended use. Their most important contribution is developing the Operational Views. After the government acquisition agency begins their program they will most likely take over the DoDAF and work out the details with the contractor(s) and external entities that include users and interfaces. They may even elect to have the contractor take over the maintenance of the DoDAF, but the government retains responsibility. There are several tools for developing a DoDAF including IBM's Rational® System Architect®. Vitech's CORE© software tool allows the engineer to capture functional and physical architectures in traditional systems engineering methods and has a DoDAF output it can generate. As a minimum a contractor's system engineers are expected to understand and work with a government generated DoDAF document for their program. They may also be involved with generating and maintaining the DoDAF at some point in the program. It is important to learn and understand the various aspects of this visualization method.

In summary, for the engineer who is working on government acquisition programs, there are specific rules and requirements he/she must follow to ensure compliance with the DoDI 5000.02 development process. The process and documents the government requires are in general good systems engineering practices. The biggest challenge to the systems engineer on a government program is to follow the right systems processes under very challenging schedules. In this aspect, working government programs is no different than commercial.

Exercises:

Answer the following questions as a review of Chapter 1.

1. Name four of the nine elements presented as being key to effective product development.
2. Rate your organization's practices against each of the nine elements for effective product development.
3. Why is systems engineering necessary for successful development of commercial products?
4. Why is it important for contractors doing business with the US Department of Defense to understand DoDAF?

2 ENGINEERING METHODS FOR PRODUCT DEVELOPMENT

2.0 Evolution of Systems Engineering Methods

Engineering design methods evolved as the complexity of products increased due to the opportunities for more complex products that arose as new technologies and production methods became available. The developments of electronics in the mid-20th century and low cost digital processors and memory in the late 20th century were major drivers for increasing product complexity and causing the need for more and more sophisticated engineering design methodologies. The concurrent growth of more efficient production methodologies enabled the more complex products to be affordable. A brief review of this history and some of the noted chief engineers is helpful in understanding the modern methodologies. (Today we would call these individuals system engineers but that term didn't appear until the late 1950s or early 1960s.)

2.1 Chief Engineer Era

Before the middle of the 20th century products were relatively simple and large design margins were often used, which facilitated simple engineering methodologies. The principle engineering method was a chief engineer plus some assistants. This model is called the "craftsman" model because it's like the methods an individual craftsman uses in developing a new product. Let's look at one example. Gordon M. Buehrig was a noted automobile designer from the 1930s to the 1950s. In 1935 the Cord Automobile Company was in trouble and needed a new design. Buehrig and his team developed the Cord 810, shown in Figure 2-1, in about six months, if my memory is correct.

The Cord 810 was the first American front-wheel drive car with independent front suspension. The design had numerous innovations including hidden door hinges; rear hinged hood, disappearing headlights, concealed fuel filler door, semi-automatic transmission in front of the engine, variable-speed windshield wipers and a radio as a standard feature. The Cord 810 was a sensation at the New York Automobile Show in November of 1935 with Cord having rushed to build the 100 cars needed to qualify for entry. This rapid design and initial production temporarily saved the company. The design was recognized for its innovation by the Museum of Modern Art in 1951.

Figure 2-1 The innovative Cord 810 developed by chief designer Gordon Buehrig and his small team in just a few months in 1935. Photo Courtesy of Auburn Cord Duesenberg Automobile Museum, Auburn, Indiana

Can you imagine a modern car company designing and producing 100 new cars with many new and even radical design innovations in about six months? Visiting Buehrig and his team's offices preserved in the Auburn Cord Duesenberg Automobile Museum provides some clues to how it was accomplished. The small team occupied a room with just enough space for the designers and their drafting boards. Buehrig's office was adjacent with only a glass window and doorway separating the two spaces. He was able to see every team member with a single glance and he could be at the side of anyone within a few seconds. This co-location enabled Buehrig to see all of the design information almost instantly and to ask questions and get answers almost instantly. Thus within the design team there was almost no information latency, that is the time between when information is available and when it gets to the individuals that need it for the next steps in their work.

The team was located in the same building as the top management of Cord and the factory was next to the office space and reachable within a minute. If Buehrig's team needed information about production capabilities or a decision from top management they could get the answers needed within minutes or a few hours at most. Of course I don't know how it really worked in practice but the physical space where this team worked was organized for minimizing information latency and this is the lesson we should take from this story.

A second important contributing factor is the fact that cars were much simpler in the 1930s and a chief designer with Buehrig's talent could understand all the important design criteria and the merits of the design options available to his team, as well as the production capabilities of the factory. This enabled the team to be limited to a few designer/draftsmen and facilitated the speed of information flow. Today's complex automobiles cannot be developed with such small teams.

Was the rapid design of the Cord 810 an anomaly? No, I could recite many such cases, particularly during WW II and immediately afterward. Many of my stories would be about airplane developments since I worked in the aerospace industry. Most people have heard of Lockheed's famous chief designer Kelly Johnson and the Skunk Works. Skunk Works has become the buzz word for organizations and facilities intended for rapid product development. Johnson's methods are well worth studying and applying today, although it is necessary to take into account the impact of modern computer technology on information flow.

Rather than recite more stories of famous chief engineers I will discuss some principles from a company manual that I had the opportunity to read nearly 30 years ago.

The company had a grand history of successful and rapid airplane developments. Someone incorporated the "lessons learned" into a manual that was both a history and a compendium of principles for rapid development. The airplane factories from WW II that I have visited had long linear production lines with shops and offices along each side and on mezzanines overlooking the production lines. Thus for a development project where the goal was to build the first flight model it was possible to locate the engineering team in a common space within sight of the area where the first plane was being assembled. I'll recite only two of the lessons learned from the manual I read. One was that the engineering team was to be located in sight of the plane and where they could get up and in a few seconds inspect the hardware they were developing. Second, the team was to be co-located in a high walled space so that the progress of each sub-team could be plotted on a large schedule chart and posted high on a wall where everyone could see the progress of each team.

Again we see the attention paid to minimizing information latency. An engineer that had questions about some part of the plane could quickly look at the hardware, if that would yield the answer. Everyone could see each other and if an engineer needed information from another engineer or even from another team it took only seconds to walk to the work space of the engineer with the answers. Posting the progress of each design team so that it was visible to all the teams had two positive impacts. First, there was the peer pressure to not be the team that was holding up progress. Experience had shown that a team that was behind would find ways to catch up with little or no management intervention. Second, the chief designer could concentrate on technical issues rather than managing schedule issues.

From the automobile design example cited above and the principles used for rapid development of airplane designs we see that the pressure for getting products to market fast are not unique to the 21st century. Twentieth century engineering methodologies adapted to this pressure by organizing and locating engineering teams to minimize information latency. The model of a chief designer plus some assistants worked well because the products were relatively simple and often large design margins were acceptable. The low complexity of the products enabled the chief designer to understand all that was necessary to achieve a balanced design that was manufacturable at acceptable cost. Thus the design process was what we now call a concurrent process; the chief designer knew the manufacturing capabilities and designed to conform to these capabilities.

2.2 Introduction of Systems Engineering

Following WW II products like cars and airplanes became more complex and the chief engineers could no longer understand everything necessary to achieve a balanced design. This lead to the introduction of specialists in the design of various subsystems or analytical specialties like stress analysis and aerodynamics. For example, instead of just mechanical engineers, specialists in structural analysis and thermal analysis developed to support the designers. Soon it became impossible for the chief engineer to understand how the engineering data from all the various specialists impacted the overall design. That is when systems engineering emerged and a team of systems engineers replaced the chief designer. It is easy to see that as more and more specialists were needed the engineering design teams grew larger and larger. It became difficult or impossible to co-locate such large teams and the communications needed to keep the systems engineers and the specialists teams current became complex and slow compared to the previous methods. As a result information latency grew along with the complexity of the products and the size of the engineering teams needed to develop the products. This lead to longer and longer development cycle times. The design process became a sequential process with periodic design reviews to evaluate status and design integrity. This evolution took place roughly between 1950 and 1980 and enabled very complex products to be developed.

The large engineering teams of specialists and systems engineers developed some fantastic products such as the Boeing 747, the Space Shuttle and modern fighter planes. However, the product development cycles were often years long and exhibited undesirable cost and schedule overruns. In addition product quality was sometimes lacking. Examination of this process reveals four fundamental flaws:

1. Too much time and money passes before design errors are found at design reviews
2. Integration of knowledge required for product development is greatly complicated by the separation of specialists into functional groups
3. There are attempts to specify cost, schedule and quality; however, only two are independent.
4. Products often reach manufacturing before manufacturing processes are fully developed.

New methods have been introduced to eliminate or at least mitigate these flaws.

2.3 Introduction of Concurrent Engineering

Many new methods emerged in the 1980s and 1990s to address the flaws in the sequential process. This book isn't about systems science or a history of all new methods; rather it covers a few methods that the authors believe are fundamental. An important new method emphasized concurrence. Manufacturing and test processes are developed concurrently with the product design. Systems engineers and design engineers work together from the start of a project; replacing the previous approach of sequential development of specifications and design. Concurrency is facilitated by early involvement of manufacturing, reliability, procurement, vendors etc. in the design process. The new methods emphasize preventing problems rather than fixing problems by incorporating verification and validation at each step.

Note the IEEE Std 1220-1998 document *IEEE Standard for Application and Management of the Systems Engineering Process*[1-3] defines concurrent engineering as "The simultaneous engineering of products and life cycle processes to ensure usability, producibility, and supportability, and to control life cycle and total ownership costs." Concurrent engineering is sometimes called integrated product development (IPD) or integrated product and process development (IPPD). In this material the IEEE definition is used for the product design process and IPD for the way the product development personnel are organized to conduct concurrent engineering. There are no universal agreements for these definitions, or other systems engineering nomenclature, so the reader is again cautioned to be careful in interpreting systems engineering terms from different sources.

Accompanying the development of concurrent engineering in the 1980s and 1990s were methodologies for improving product quality. Two such methodologies imported from Japan are Quality Functional Deployment (QFD) and Taguchi Design of Experiments. The term robust design emerged to describe engineering using these quality improvement methodologies. Clausing[2-1], Phadke[2-2] and Huthwaite[2-3] offer excellent books on these subjects.

2.4 Further Improvements in Design Methods

Engineering is inherently an iterative process due to the fact that all engineers make mistakes in their work. Engineering work involves making hundreds of decisions in each task and often involves complex analysis. It is almost impossible to do such work without making some mistakes on the first pass. A result of this fact is that design quality and design cost are linked by the number of design iterations performed. This is shown schematically in figure 2-2 for a product of some specific complexity.

If a design of quality q is desired then n iterations are needed and the cost will be c for a given set of design tools and methodologies. Notice that the schedule, cost and quality cannot all be specified. If a certain schedule is required then either the cost or the quality can also be specified, but not both. All three are not independent variables.

Also note that this simple diagram shows that if an organization wishes to lower costs, or shorten the design cycle, for a given quality it is necessary to introduce new tools, new methods or both.

2.4.1 Achieving Reduced Product Development Cycle Time - Experience has shown that product development cycle time can be reduced using more effective engineering methods as shown in Table 2-1. Although the factors listed in the table are approximations and dependent on many specifics up to about a factor of ten reduction in cycle time is available by learning and properly using modern methods. These modern methods for systems engineering and their proper use are the subject of this book.

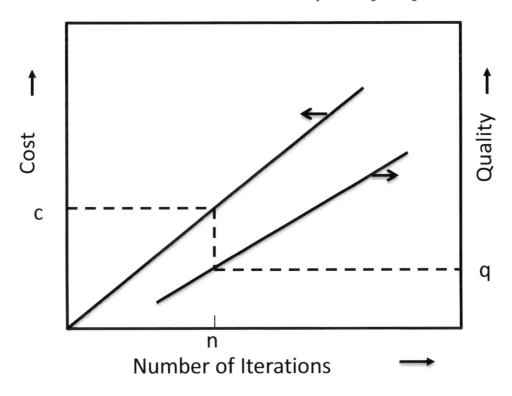

Figure 2-2 Design quality is linked to design cost by the number of design iterations performed by relationships that are dependent on the product complexity and the engineering methods used.

Approach to Reducing Cycle Time	Time Reduction Factor	Cumulative Reduction Factor
Eliminate Design Defects	2x	2x
Use More Effective Processes	2x	4x
Eliminate Redundant Effort	2.5x	10x

Table 2-1 Experience demonstrates that cycle time reduction can be significantly reduced by using modern methods.

2.4.1.1 Eliminating Design Defects - Techniques for eliminating defects include peer reviews, using graphical models in place of text and disciplined validation and verification throughout the design cycle. The benefits of using graphical models are discussed in Chapter 3 and validation and verification processes are covered in a later chapter.

Rather than depend on design reviews of major segments of work to find problems **peer reviews** are held as soon as an individual or small team completes a task. Self-checks catch most mistakes but the best way to find errors after performing self-checks is for the engineer to review the work in detail with highly experienced peer engineers. Experienced peers provide an independent and often different experience background that helps reveal errors.

Catching errors at peer reviews allows the errors to be corrected before the results of the work are incorporated in following tasks. Thus little rework is involved in fixing errors found during peer reviews. In contrast errors found at periodic design reviews involve considerable rework because often the erroneous work has been incorporated in several following tasks that must also be reworked. Extensive rework due to errors found at design reviews on complex product developments is a significant contributor to cost and schedule overruns.

If teams are disciplined in holding peer reviews then the design reviews can become more of a status and decision gate review to communicate to managers and customers the state of the design and to check that all required actions are complete per a design review checklist. Further, if any reviewers wish to go into detail on some design point the documentation that engineers have previously used in the peer reviews can serve for more detailed discussions. This reduces the time and costs involved in preparing for design reviews.

2.4.1.2 Using More Effective Processes - A major factor in achieving cycle time reduction by eliminating defects is whether the design team uses practices that prevent design problems rather than find and then fix problems. It is much cheaper and faster to prevent problems than it is to find and fix the problems. Risk management, which is discussed in detail in a later chapter, is the primary method of proactive problem prevention. Other processes that contribute to reducing cycle time include: Progressive Freeze, Comprehensive Functional Analysis, Pugh Concept Selection, Quality Function Deployment (QFD), Taguchi Design of Experiments and Spiral Development. Functional Analysis, QFD and Pugh Concept Selection are covered in later chapters. Taguchi Design of Experiments, a specialized methodology used extensively to improve or trouble shoot manufacturing processes, is also an effective methodology for improving designs and can be applied at any point in the design cycle. The reader is referred to the books of Don Clausing[2-1] and Madhav Phadke[2-2] cited above for instruction in this valuable methodology.

Spiral Development - Instead of designing the ultimately desired and often highly complex system in one step spiral development sets limited objectives for the first development effort, i.e., the first spiral. Lessons learned from the first spiral are then incorporated in the second spiral. Just like peer reviews spiral development recognizes the inherent iterative nature of product design engineering. The nature of spiral development can be understood from the diagram in Figure 2-2. A highly complex product would have a diagram with a very steep curve relating design iterations to cost. By developing a lower complexity product in the first spiral the relationship is less steep; that is the cost for a given number of iterations is lower. Similarly, the quality is higher for a given number of iterations for the lower complexity product. Thus the desired quality of the first spiral can be achieved for lower cost than for a more complex design. The same holds for later spirals. The lessons learned on any spiral are applied to later spirals, resulting in more favorable relationships between cost, schedule and quality than if the entire complex design is done in one spiral.

Progressive Freeze - Progressive freeze of design decisions is another technique introduced to help manage schedules during the development of complex products. It involves freezing a design detail as soon as it becomes likely that further work won't change it and before the next major design review. This enables more detailed design work to begin ahead of the design review intended to gate such detailed design. This provides schedule and staffing flexibility that can help avoid delays due to the unavailability of skilled personnel.

2.4.1.3 Eliminating Redundant Effort – Table 2-1 shows that eliminating redundant effort offers the largest reduction in cycle time. Two techniques for helping eliminate redundant effort discussed in later chapters of this book are **decision management** and the use of **pattern based systems engineering** (PBSE). The goal of decision management is not just to manage the decision process but to make the process robust enough that decisions can be reused on future developments. PBSE, introduced in Chapter 5, is a technique that uses graphical models and textual templates to reduce the systems engineering effort. The graphical models are constructed for a family of products so that the model for any particular product is developed by deleting elements of the pattern not applicable to the particular product being designed. This reduces the time to develop the model by as much as a factor of ten and is less prone to errors. Decision management and PBSE are methods for implementing one of the elements of effective product development listed in Chapter 1: never having to redo something that was done correctly on a previous development program.

2.5 Review of Fundamental Principles

Concurrent engineering, IPD or IPPD, spiral development, and robust design represent the typical engineering processes for product development in use at the turn of the 21st century and are still widely used. However, the engineering design process has not stopped evolving to match the ever increasing complexity of products. The introduction of newer methods is the intended focus of this book. But before introducing these newer methods it is worthwhile to step back and review some fundamentals of product design methods.

Examination of the "lessons learned" from over 50 years of evolving product design methods suggest that there are some basic principles that should be considered when examining new methods or tools. Some are evident in the examples given in this chapter and others derive from the experience of successful systems engineering teams. These can be summarized in **12 principles** that underlie modern product development methods:

1. The best model for a simple product development is a "Craftsman" model, i.e. a Chief Engineer with total project visibility and responsibility.
2. Complex products should be decomposed into simpler pieces so that the craftsman model can be applied to each piece.
3. All activities in a product's development must become one integrated effort. (Otherwise, #2 cannot be implemented successfully.)
4. Trade-off within single disciplines must be subordinated to trade-offs across disciplines. (Necessary to achieve a "balanced" design.)
5. Authority must be shared among members of a collaborative, multi-disciplined team.
6. Product complexity will continue to increase, leading to the need for more engineering specialization. (And the need for better systems engineering tools)
7. The complexity of system engineering tools should be matched to the complexity of the product and to the complexity of the design element within the system hierarchy; i.e. system, subsystem, etc. (Note that a process must be fully understood before choosing and implementing tools for that process.)
8. Continuous Process Improvement is essential to compete successfully in a global market.
9. A competitive advantage is realized from technology and tools that provide all team members complete visibility of evolving designs, support in identifying and resolving conflicts, and the ability to equally influence decisions.
10. Modern methods are essential to shorting the time to prevent; find and fix design errors and thereby facilitate achieving minimum product development cost and schedule.
11. Functional managers must collaborate, share responsibility in measuring team performance, and foster unity of purpose within and among teams.
12. Product development schedules must account for resource contentions within the team and among teams.

The reason the **Craftsman model** is the best for simple products is that all the information necessary to guide the design is available in the mind of the experienced leader or chief engineer. This is best because it reduces information latency to an absolute minimum. If information latency is minimized then the conditions for minimizing cost and schedule are in place. This means that cost and schedule will be minimized assuming the other principles are also fulfilled. (The principles could be summarized in a shorter list if the principle of minimizing information latency is included. Here this fundamental principle is decomposed into subordinate principles because it makes it easier for systems engineers and product development managers to see how the principle should be implemented on their projects.)

The second principle derives from the first and tells us part of how a product development team should be organized to minimize information latency. It says to decompose the product into pieces (subsystems, assemblies, etc.) to the point that a single lead engineer can be knowledgeable enough to guide the development of the piece. This lead engineer must be supported by the needed specialty engineers and must work with other team leaders to achieve a balanced design but the individual should be capable of being responsible for all design decisions on the design element assigned. How all these small teams are organized into an overall integrated team is the subject of principle 3 and the core organizational issue of IPD.

Principle 4 says that one of the responsibilities of systems engineering is to manage the tradeoffs between different disciplines (and teams) so that a balanced design is achieved. For example, compromises are often needed between electrical design and thermal design or between mechanical design and optical design. If the systems engineers allow one discipline to dominate then that discipline will make their portion of the design optimum at the expense of other disciplines; resulting in an unbalanced design which is usually not robust to the variability of use. Principle 5 is a necessary condition for successfully implementing principle 4. Principles 6 should be self-evident and principles 7 and 8 follow from 6.

Principle 9 is derived from examining the question of how we achieve the benefits of the craftsman model for the development of modern complex products with many engineering specialties. For a number of years I used this principle in training systems engineers and had to admit that I did not know how to achieve the goals listed in this principle. New methods emerged in the late 1990s that now enable teams to achieve the goals of principle 9. One important new method, called **Integrated Concurrent Engineering** (ICE), is discussed in chapter 12.

Principle 10 says that teams must use methods like peer reviews, progressive freeze, quality function deployment and risk management to achieve minimum cost and schedule. This list is meant to be illustrative and is definitely not a complete list of modern methods and communication tools needed for effective systems engineering and product development. None of these methods are difficult but teams must exhibit good discipline to maintain effective use of these tools and inexperienced managers sometimes fail to maintain the necessary discipline.

Principles 11 and 12 are "housekeeping" principles and perhaps could be left off this list but are added as reminders.

Exercises

Before proceeding to Chapter 3 take a few minutes and consider the following questions:

1. What is the "Craftsman" model of product development?
2. What replaced the Craftsman model and why did it have to be replaced?
3. Review the product development processes in your organization and rate them against the list of 12 principles stated above on a scale of 1 to 5, with 5 being full satisfaction of the principle. If your organization achieves a score of 50 or above it's likely that it's competitive with most of its competitors. If the score is below 50 then there is work to do even before trying to implement the 21st century methods that will be introduced later.
4. If your organization is required to hold extensive design reviews that take several days and many engineering hours to prepare for think about how peer reviews could be used to reduce the effort needed to prepare for design reviews.

5. Describe the end to end product development process in simple terms your grandmother could understand (assuming she is not a systems engineer) if she asked "what are the steps in product development"?

3 INTRODUCTION TO SYSTEMS ENGINEERING

3.0 Overview

Systems engineering is defined in many texts. Rather than repeat the standard definitions at this time it is helpful to define system development in the simplest possible manner. The reason is that having very simple definitions helps us to consider things from a fundamental point of view, which makes it easier to come up with new insights. New insights are needed in order to understand new methods and, eventually, to develop improved methods. This chapter defines four fundamental tasks that systems engineers perform, provides definitions of systems engineering, compares the rolls of systems and design engineers and shows how systems engineers fit in a modern product development organization. The four systems engineering tasks are deliberately defined before defining systems engineering. Although this seems out of order it helps take a fresh look at the systems engineering process. To set the stage for defining the four tasks of systems engineering we look at the overall product development process or development cycle as it is usually called.

3.1 The Product Development Cycle

In the exercises at the end of Chapter 2 the reader was asked to define the product development cycle in simple terms a grandmother would understand, assuming the grandmother is not a systems engineer. The purpose of this question is to make the reader think about the fundamental tasks that comprise product development. Sometimes people get into the detail of their work to the point they lose sight of the fundamentals of what they are doing. Figure 3-1 illustrates what the reader might have said.

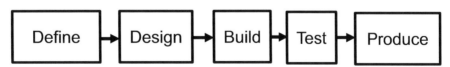

Figure 3-1 Product development consists of five tasks, or phases, at a fundamental level.

Product development organizations typically define the product development cycle in a much more complex diagram, or set of diagrams, that describe the tasks to be performed in much more detail than shown in Figure 3-1. However, it is important to step back and look at the fundamental tasks, or phases if you prefer, in as simple of terms as possible if we intend to improve the overall process. Starting with the simple flow chart in Figure 3-1 we define the work that systems engineering encompasses, again in simple terms, because our objective is to seek better methods for systems engineering. Note that Figure 3-1 leaves out the object of the verbs. This forces us to think about what it is that we are defining, designing, etc. We can add some of the top level inputs and outputs of each step in Figure 3-1 as we think about what it is that we are defining, designing, etc. This is shown in Figure 3-2.

The tasks shown are performed by a variety of managers, technicians, and engineers; systems engineers, design engineers, quality engineers, manufacturing engineers, test engineers, etc. We seek to define just the tasks performed by systems engineers. Actually systems engineers are involved in all five tasks to some extent. Typically systems engineers are responsible for the specifications and plans that are produced in the "Define" and "Test" tasks and contribute to or oversee scheduling and the work in the other three tasks. With these roles in mind we can think about the tasks the systems engineers perform. I think it adds clarity to define the tasks first and then define the processes used to carry out these tasks.

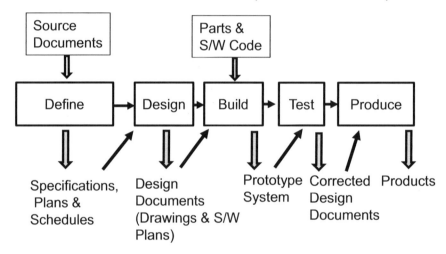

Figure 3-2 The outputs of a fundamental product development task are the inputs of the next task

3.2 Four Fundamental Tasks for Systems Engineers

A simple but useful terminology for defining systems engineering work is shown in Figure 3-3.

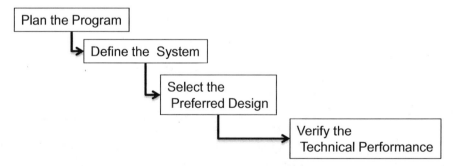

Figure 3-3 Four fundamental tasks performed by systems engineers during the product development cycle.

The four tasks listed in Figure 3-3 are not the only way to define systems engineering tasks, or are these a complete description, but these four tasks help us understand the fundamentals, if we allow for some flexibility in our definitions. This becomes apparent as we examine in more detail the four tasks in later chapters.

Notice that the first three systems engineering tasks are shown as overlapping in time rather than being sequential in time. This is because these tasks should overlap in time. This is a good point to introduce one complexity in the fundamental task list shown in Figure 3-1, then we can return to defining systems engineering.

3.3 Progressive Freeze of Product Development

The product development tasks should not be rigidly sequential but should allow for some overlap in time, as indicated in Figure 3-4.

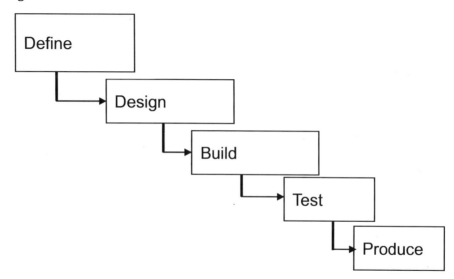

Figure 3-4   The fundamental product development tasks should be allowed to overlap in time.

Often major product development tasks, or phases, are gated by milestone reviews. This implies that all work must be completed on one phase of the product development before work can start on the next phase. It is not good practice to demand rigid adherence to the gating process. A better approach is to allow some work to begin as soon as the information needed for initiating the task is available. This is called progressive freeze, as discussed in the previous chapter, and it adds flexibility to the work that offers benefits that outweigh the small additional risks incurred.

Progressive freeze is practiced by freezing an element of the engineering work, when the risk of freezing the work is acceptable and allowing work on the next phase to begin before the next milestone review. Engineering judgment is used to determine when the risk is acceptable that the continuing work on other elements of the development work won't change the results of the frozen work element.

A couple of examples help clarify the progressive freeze concept. Consider a radio frequency (RF) sensor. Freeze the antenna architecture when the gain, antenna configuration and interfaces are determined; initiate the antenna design in parallel with other subsystem architecture trades. Hold a system design review (SDR), if required, when all subsystem architectures are frozen. Second, consider an infrared sensor. The telescope architecture can be frozen when the aperture form, the optical subsystem f/# and interface are determined. The telescope design can then proceed while the focal plane, electronics and cooling architecture trades continue.

There are several benefits from beginning the design early for the RF antenna or the telescope in these two examples. The biggest is the flexibility in scheduling the skilled antenna and telescope designers. Critical skills such as these are often in high demand on several programs. By enabling early starts of the design work involving critical skills it increases the total time period for the design work to be accomplished without impacting the program schedule and helps avoid program delays due to the unavailability of critical skills.

A second benefit is that progressive freeze facilitates sequencing design decisions throughout a design phase. Product development involves thousands of decisions in each phase of the work. Delaying decisions until just before major design reviews results in such decision management complexity that the quality of the work can be threatened. Making decisions as soon as risk allows spreads the decisions throughout a development phase and makes decision management easier. We will return to this issue in a later chapter where we explore design decision management in more detail.

Another benefit is that should a problem arise in the design that impacts other architecture the architecture changes can be made easily if these trades are still underway. For example, suppose the RF antenna designer discovers that the desired gain margin cannot be achieved within the constraints defined in the architecture trades. If the other trades are still underway it is much easier to decide how to accommodate the problem. It may be that a larger envelope is available or it may be that the lower gain is acceptable with tighter control of noise figure in the amplifier electronics. Finally, allowing more time for the designers to work on a design element may result in a better design. Sometimes engineers need to set work aside for a short period while they mull over a difficult design issue. Sometimes unforeseen difficulties arise in design and having extra time to work out these difficulties helps avoid program delays.

3.4 Purposes of Systems Engineering

Now that we have taken what is hopefully a fresh look at what systems engineers do let's back up and look at some traditional definitions of a system and systems engineering. I have used these definitions for so long that I have forgotten their original sources and I apologize. There are almost as many definitions of a system and of systems engineering as there are systems engineers so I prefer not to put much emphasis on these definitions but include them for completeness and to make a critical point.

One source says a system is a collection of objects working together to produce something greater. A system can be tangible, like a rocket or a communications network, or intangible like a computer software program. A system is composed of unique elements and the relationships between these elements. A system has the further property that it can be unbounded; each system is invariably a part of a still larger system.

Eberhardt Rechtin, in his book *Systems Architecting: Creating & Building Complex Systems* [3-1], says the essence of systems engineering is structuring. Structuring means bringing form to function or converting the partially formed ideas of a customer into a workable model. The key techniques are balancing the needs, fitting the interfaces and compromising among the extremes. Whereas this definition is essentially correct it misses an important point. Much of systems engineering is defining the relationships between a system and its environment, including other systems, and between the different elements of the system.

Others describe systems engineering as both a technical and management process. One source says "it is a discipline that ties together all aspects of a program to assure that individual parts, assemblies, subsystems, support equipment, and associated operational equipment will effectively function as intended in the operational environment". An interesting definition is from a U.S. Navy web site. It says systems engineering "is a logical sequence of highly interactive and iterative activities and decisions transforming operational needs into a description of system performance parameters as well as a preferred system configuration." This definition is amusing because it is exactly correct in saying systems engineering is highly interactive and iterative but the reality is that it is logical only to a highly experienced system engineer. Something that is highly interactive and iterative hardly appears logical to someone not intimately familiar with it.

Comparing these definitions with the four tasks for systems engineers defined in Figure 3-3 shows that the definitions do not capture all that is implied in the four tasks. From the definitions would a young engineer realize that a system engineer has a role in planning and scheduling a development program; or in verifying the performance of a system design? If this chapter started with the text definitions of a system and systems engineering would a logical path to the four systems engineering tasks in Figure 3-3 have been easily found? If just the five text words had been used to define the product development cycle instead of the labeled graphical model would it have raised the question of whether the four tasks are rigidly sequential or allowed to overlap in time? Perhaps, but not as easily as it is starting with the graphical model of product development as five simple sequential tasks or phases as shown in Figure 3-1.

3.5 Graphical Models vs. Textual Models

There is a profound principle in the last paragraph above. Textual descriptions of complex ideas are often incomplete and very often lead to different readers having different understandings of the ideas. Labeled graphical models of complex ideas are less ambiguous. This is a critical point for systems engineers to remember and use in their work. Look back at Figure 3-2. The thin solid arrows indicate that the outputs of a task are the inputs to the next task. In the development of complex systems each task typically involves different teams. Therefore it is critical that the outputs of one task be easily understood and unambiguous to the teams working on the following tasks. Labeled graphical models are much easier to understand and less ambiguous than textual descriptions for most engineers.

Think back to the Craftsman model of product development defined earlier that was successfully used for centuries for simple products. The output documentation from the chief engineer and his/her team was engineering drawings. The workers and managers in the factories responsible for producing the products could clearly understand and use these engineering drawings. Imagine if the chief engineer and his team submitted textual descriptions of their product design to the factories.

Another example of a labeled graphical model being superior to textual descriptions is maps. Imagine going on a driving vacation through your country using just a textual description of the location of towns and the roads connecting them. Just keeping track of where you are with respect to the textual descriptions would become a time consuming task. This is why maps became the standard for describing geography centuries ago.

One of the reasons development of complex systems often failed to achieve the planned schedule and budget after systems engineering was introduced in the 1950s and 1960s to address the complexity of modern systems was that some of the documentation produced by systems engineers and passed to design engineers was poor textual descriptions rather than labeled graphical models accompanying carefully written textual documentation. Unfortunately this practice continued when concurrent engineering was introduced and continues today for many organizations. One of the primary objectives of this book is to convince systems engineers to use labeled graphical models wherever possible and avoid textual documentation unless required by customers. This is one of the keys to achieving faster and lower cost systems engineering.

3.6 System Engineering and Design Engineering

Accompanying the introduction of systems engineering was the practice of distinguishing between systems engineers and design engineers. It was natural to label the engineers doing the systems engineering work as system engineers. Since functional engineering organizations were typically broken down into mechanical departments, electrical departments or some similar differentiation it was natural to set up systems engineering departments for the systems engineers. This is unfortunate because all engineers do some systems engineering work and most engineers do some design engineering work. It is true that systems engineering has become a specialty just like other specialties of engineering. We must keep the various specialties but perhaps we could drop the term design engineer. This might help remove some of the undesirable barriers that creep into our organizations that inhibit effective communications. However, using the term design engineer helps define rolls of the various specialties. Let's use the diagram from Figure 3-4 to help define the traditional roles and responsibilities of systems engineers and design engineers. The result is shown in Figure 3-5.

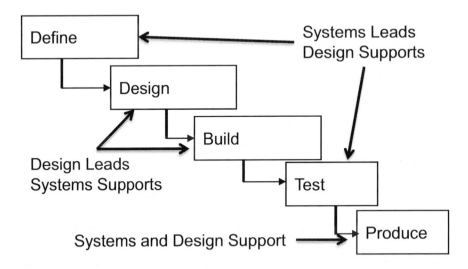

Figure 3-5 The traditional roles of system engineers and design engineers involve responsibilities within each development phase.

If we didn't want to use the term design engineer we could say specialty engineers or "other specialties". Although that would be awkward it is more accurate because saying design engineers leaves out other specialties like quality engineers. For convenience in this work the term design engineer is assumed to include any specialty engineer other than a systems engineer. From Figure 3-5 and the description of what takes place during each task or phase we can further define the traditional differentiation between systems engineers and design engineers as follows:

- Systems Engineers Determine
    - What's to be built
    - How it's supposed to function
    - Whether it meets its requirements
- Design Engineers
    - Determine how it's to be built
    - Make it operate (integration & test)

It is critical to understand that both systems engineers and design engineers are involved in all five phases as shown in Figure 3-5. To further define the role of systems engineers it helps to look at how they are integrated into a typical organization.

3.7 System Engineers in the Organization

Let's assume a product development organization practices concurrent engineering and organizes in a set of interlocking teams. The top level team is often called the core team. A core team for a medium to large development program is composed of representatives from the major contributing functional organizations in the enterprise and might look like Figure 3-6.

A circle structure rather than a tree is used for the organization because each member of the core team is functionally independent and has responsibilities both to the product development or core team and to his/her functional organization. This means for example that the contracts representative performs within the team in the best interests of the product development program but is also constrained to follow the policies and procedures the parent organization defines for the contracts function. Similar rules apply to the other members of the core team. Each member of the core team is a leader of a support team, e.g. the engineering member of the core team is the lead engineer for the engineering team.

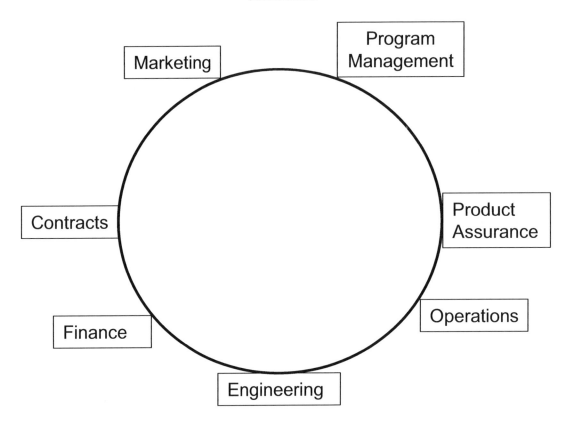

Figure 3-6 A typical core team for a medium to large product development program has representatives from all the contributing functions.

Now we can look at how the engineering work is organized and how systems engineers fit within with the engineering team. An example is shown in Figure 3-7.

In Figure 3-7 SE stands for a systems engineer and DE stands for an engineering specialist other than a systems engineer. The engineering leader is called the project engineer, as in this figure, or chief engineer, or lead engineer. The leader of the System Engineering and Integration Team (SEIT) is usually called the lead systems engineer. The engineering IPT is thus composed of a number of subordinate IPTs, here labeled A through N, which are responsible for subsystems or for specialty functions like test or software. The systems engineers shown under the subordinate teams in Figure 3-7, or the subordinate team leaders themselves, are members of the SEIT as well as their respective IPTs. If the system being developed is large and complex than there are likely many more layers of IPTs. Think of each of the DE blocks being an IPT with the SE block being a lower level SEIT.

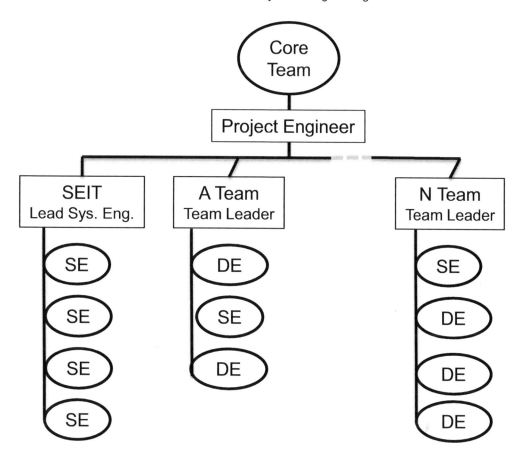

Figure 3-7 The engineering integrated product team (IPT) is typically organized with a Systems Engineering and Integration Team (SEIT) and a number of teams responsible for the various subsystems or engineering functions such as software or test.

Not explicitly shown in Figure 3-7 is how members of the operations team, the product assurance team and key suppliers support the engineering team and vice versa. In some cases it may be sufficient to only have the structure shown in Figures 3-7 and 3-8 but in most cases an operations engineer and a quality engineer should be a member of the SEIT and a member of the SEIT should be a member of the operations IPT and possibly the product assurance IPT. Usually an engineer knowledgeable about a supplier's product is responsible for coordinating with the supplier and bringing the expertise of the supplier into the design process at each phase of the design work. If a supplier is a teammate or strategic partner in the development then supplier engineers can be members of the engineering IPT.

The principle to be followed in organizing the engineering team is to break down work so that at the lowest level IPT the craftsman model applies. The work assigned to each of the lowest level IPTs should be sufficiently low in complexity to be thoroughly understood by an experienced team leader. Thus the team leader is a "chief engineer" for the work assigned to his/her team.

The nesting of interlocking IPTs can continue as necessary to handle the size and complexity of any large system. This nesting of teams reduces the time non managers spend in communications meetings. In the three tier example shown in Figures 3-7 only the lead engineer, here called the project engineer, needs to attend core team meetings although sometimes the lead systems engineer may attend. Engineering coordination meetings are held at two levels. The project engineer meets with all the team leaders and then the team leaders meet with their respective teams as necessary to flow down important information. This approach limits the amount of time workers spend in coordination meetings and limits the meeting topics to just those items of interest to each team. Few things are more frustrating to working engineers than having to sit through meetings on topics that are of little or no interest to them; plus it's a waste of time and money to have working engineers sitting in nonproductive meetings.

Product development programs usually have documentation called a work breakdown structure (WBS) that assigns code numbers to various elements and phases of the work so that budgets can be generated and managed. The WBS can map to the organization to facilitate cost management although this is not a strict constraint. If the IPT is organized according to a SEIT and subsystems then the WBS can map to both the organization and to elements of the system. Organizing WBS's in this manner facilitates developing data bases of costs associated with the different subsystems and development phases thereby making future cost estimating easier.

Figure 3-7 explicitly shows that there is systems engineering specialty work to be done in each of the subordinate IPTs by including a SE on each subordinate IPT. There is systems engineering work to be done even if the person doing the work is called a design engineer rather than a systems engineer. Some design engineers think that they don't have to do systems engineering work, which just isn't true. They do systems engineering work even if they don't use all of the tools used by systems engineering specialists. Since a system is always part of a larger system there is systems engineering work in all product development at all levels of "systems". Even though an engineer is developing the design for an assembly or component of a system there is systems engineering work required. Of course this systems engineering work must be tailored to the level of complexity of the design element.

Systems engineers can perform a variety of roles on IPTs, from lead systems engineer to contributing engineer or analyst on one or more other teams. Since the IPTs are multidisciplinary, it is critical that the systems engineers understand their roles and responsibilities on each team. A good practice is to hold roles and responsibilities meetings with all members of an IPT as soon as the team is staffed.

The engineering team for a small product development program can be simplified from that shown in Figure 3-7. An example of an engineering IPT for a small project is shown in Figure 3-8.

Figure 3-8 shows an engineering IPT with just seven engineers; a project engineer (PE), a system engineer (SE), an electrical engineer (EE), a mechanical engineer (ME), a quality engineer (QE), an operations engineer (OE) and a test engineer (TE). This organization might be sufficient for developing the design of a small electrical or electro-mechanical system. Note that this organization clearly illustrates the multidisciplinary nature of IPTs and is what might also be used on one of the subordinate IPTs of a large development program.

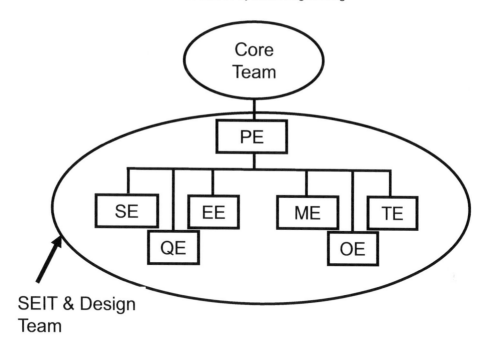

Figure 3-8 The engineering IPT for a small development program can be simple with the engineers having responsibility for both systems engineering and design engineering.

Exercises:

Answer the following questions as a review of this chapter:

1. List four responsibilities of the systems engineers on a product development program.
2. Define "progressive freeze" and list three benefits of using progressive freeze.
3. Why are labeled graphical models better than textual documents for communicating engineering data?
4. Why do design engineers have to do systems engineering in their work?
5. Why are both systems engineers and design engineers involved in all phases of product development?
6. What are a SEIT and an IPT?

4 PROCESSES AND TOOLS FOR PLANNING A PROGRAM

4.0 Introduction

The first activity for systems engineers on a new product development program is to assist in planning the program. Although planning a program is the responsibility of the program leadership a product development program cannot be properly planned without some systems engineering work. Program planning involves defining the tasks to be performed, scheduling the tasks and developing the planning documentation needed to guide the work and the workers. One of the primary reasons product development programs fail to achieve the expectations of management and customers is poor planning. This chapter presents processes and tools that reduce the chances that a product development fails to meet expectations due to poor planning.

The chapter discusses the various plans and documentation making up a product development plan, scheduling the development and touches on the important program management issue of supplier relationships that involve systems engineering. This chapter does not distinguish whether systems engineering or program management, is responsible for any specific plan item. That decision is up to the development team management. Allocating budget for each task and assigned people to each task are additional planning functions that are not treated here as these activities usually don't rely on systems engineering other than for estimating the cost of the systems engineering work.

Product development programs are called projects here even though the term projects refer to both product and service developments. The planning for a new product or service development is similar so there is no reason to restrict this chapter to products; besides project is a shorter term than product development program.

4.1 Planning Documentation

Documentation is an integral part of project planning. Whereas it's unlikely that many projects fail due to poor documentation good planning documentation makes a project easier to manage and easier for the workers assigned to the project. Four types of documentation that are typically used on complex projects are discussed here. These are the **Integrated Management Plan (IMP),** the **Schedules,** the **Systems Engineering Management Plan (SEMP)** and the **System Design Document (SDD)**. The first three comprise the top level planning documentation. The SDD is an information document rather than a planning document but it should be started during the planning phase. Of course there is considerable documentation needed to back up these top level plans. A key supporting plan is the **Modeling and Simulation plan**, which must be done early so that needed models and simulations can be developed or acquired in time to support the development schedule. A modern complex system or product development has so much documentation that a plan must be developed up front for defining the needed documentation and organizing the hierarchy and relationships of this documentation. This plan is called **Information Architecture** here.

There are two ways to produce poor documents and one way to produce good documents. Poor documents are either so brief or poorly written that there is little useful information for managers overseeing the project and new workers assigned to the project or the documents are so long and detailed that no manager and few workers will ever read them. Surprisingly, most bad documents are those that are far too long to be useful. This is because the project leaders have slavishly followed advice to include everything but the kitchen sink description in these documents.

In the introduction it was stated that where ever possible labeled graphical models should be used rather than textual documents. Plans are one of the areas that require a significant amount of text. However, the goal should be to make as much of the plans as possible graphic models such as diagrams and tables. (Tables are included as graphic models here because they are less ambiguous than pure text.)

4.2 The Integrated Management Plan

A concise IMP is preferred because the IMP should be viewed as a "contract" between the project team and their senior management. Therefore the document need only address issues at a summary level. This allows the team and management to focus on the important issues without getting mired in detail. Mature organizations have standard procedures and processes for items like configuration management and product assurance. Management needs to know if any tailoring is needed for standard processes but they don't need to wade through the details of standard processes. Certainly documents like a Product Assurance Plan are necessary but hopefully the organization is mature enough that the plan for any specific project is a tailored version of a standard plan. Summarizing how the plan is tailored to the specific project is sufficient for the IMP. Any manager or worker that needs to know the details can read the detailed plan.

A good IMP is about ten pages long or less. If it is any longer it won't be read by managers who are expected to have oversight of the project or by few workers that are assigned to the project. It is quite difficult to cover all the items that are recommended for an IMP in ten pages but it is far better to put effort into a good ten page document than to put even more effort into 80 pages that are never read. A template is provided here for a ten page IMP. Many books and web sites advise far more material be included in an IMP but it's not effective to produce such tomes and it costs precious time and money to produce them. If a project leader insists that all the detail is needed then include a seven to ten page Introduction and Summary for managers because that is all they are likely to read.

Table 3-1 shown here is a model for a seven to ten page IMP. Your project may not have the same required items but this model illustrates that a useful IMP can be ten pages or less. The suggested number of pages in this table adds to ten pages if the maximum suggested amount is used for every item. The recommended topics provide managers a concise overview of a project and are sufficient to orient new workers assigned to the project. If the resulting IMP is well written then, together with the project schedules, it provides senior managers all the information they need to conduct oversight that ensures that the project is progressing adequately or that the project leaders need help. New workers do need additional information but that is better provided in the SEMP and SDD.

Topic	Number of pages
Program Overview (Milestones, Deliverables with quality level required, Fit with Strategic Goals and a one page Top Level Schedule)	1.5 to 3 pages
Technical Description of System (Intended function, hardware/software overview and any make/buy issues)	1 to 2 pages (Include a top level System Breakdown Structure (SBS) and link to more detailed breakdowns in the SEMP.)
Key Performance Characteristics	0.25 to 0.5 pages (Include links to a Specification document for those needing more detail.)
Work Breakdown Structure (WBS)	1 page of top level only (Put the detail, if needed, in an appendix or link to a separate document, e.g. the SEMP.)
Cost target, Preliminary budget & Sources of funding	0.25 page
Staffing Plan (Preliminary manning forecasts, any critical resources, & any teaming arrangements)	0.25 to 0.5 pages
Key Assumptions	0.25 to 0.5 pages
Summary of major risks and plans to mitigate	0.25 to 0.5 pages and reference a risk register in an appendix
Inter-project cross feeds and dependencies	0.25 page
Capital and facility requirements	0.25 to 0.5 page
Status of Preliminary Support Plans (Configuration Mgmt. Plan, Product Assurance Plan, Environmental Health and Safety Issues/Plans, Logistics and Field Support, etc.)	0.5 to 1 page (Which are standard and which are to be tailored. Say why tailoring is appropriate.)

Table 3-1 An example model for a seven to ten page IMP.

If an enterprise accepts the definition of the IMP as a contract between the project team management and the enterprise management then there is no reason that the IMP always be a text document. It may be acceptable that the IMP is a set of briefing charts that the team presents to the management and receives acceptance or guidance from the management at the end of the briefing. Since most briefings today use electronic data projected on a screen it is easy to include links to the more detailed plans just in an electronic text document. Including appropriate links to more detailed planning documents makes the briefing charts as complete as any text document.

There are several advantages to this approach. First, it is easier to prepare a set of briefing charts covering the topics of an IMP than it is to write a high quality document. Second, if the team can get the management to set through a briefing then there is no question about whether the management has read the IMP and there should be no question about whether the IMP is approved or not. Also, a briefing allows any ambiguities to be resolved in real time. Finally, the 80 page tome is discouraged because senior management isn't likely to sit through a briefing that long.

A last but important reminder is to prepare the IMP using an IMP from a previous project as a template and strive to make the new IMP an even better template for future IMPs. This advice isn't limited to the IMP. Where ever possible look for opportunities to build on previous work and structure work so that the result makes future work easier and faster. This approach to systems engineering is one focus of this book.

4.3 Project Schedules

There are two types of schedules defined in IEEE 1220 and DoD nomenclature. One schedule is called the **Master Schedule** in IEEE 1220 nomenclature and the **Systems Engineering Master Schedule** (SEMS) or **Integrated Master Schedule** (IMS) in DoD nomenclature. It is an event driven schedule and is not an executable schedule; it is used by the program management for communicating status and to guide the development of detailed executable schedules. The Master Schedule maps to the WBS, identifies major milestones and lists deliverables.

The second type is executable, tied to calendar dates and is called the **Calendar Schedule**. The calendar schedule is also called the **Detail Schedule** or **Systems Engineering Detail Schedule** (SEDS). This schedule is used to develop earned value metrics for government contracts. The NASA Systems Engineering Handbook refers only to one schedule; called the **Network Schedule,** which is the same as the detail or calendar schedule. The NASA Systems Engineering Handbook provides instructions for planning and developing the schedule whereas the DoD and IEEE 1220 documents for systems engineering only discuss the planning that systems engineers must do to enable schedules to be developed.

Three schedules are necessary to achieving effective project control and communication with management and customers. First there is the Master schedule as defined by IEEE and DoD. Some slight modifications to the DoD definition are desirable. First, limit the Master schedule to about two levels for most projects. The Master schedule can then include the major phases of the planned work, the management reviews that typically gate the initiation of the next phase and the deliverables, including any long lead items that are on the critical path, and other selected items that drive the schedule. This usually permits a Master schedule to be displayed on a single page, which greatly simplifies communications with senior management.

Second tie the events to calendar dates since most customers for new product developments specify delivery dates. The reality is that product development teams typically must plan a project that meets some schedule constraints. Customers may claim they want an event driven rather than calendar driven program but it's rare that they actually do. Also the team has less incentive to develop workarounds to hold schedule and cost if the project is truly managed as event driven. These constraints are clear to everyone if they are included in the Master schedule. Finally, use the term Master Schedule and not the terms Integrated Master Schedule (IMS) or SEMS or SEDS. The detail schedule, or network schedule as NASA calls it, is the integrated schedule and both the Master schedule and the detail schedule apply to more than the systems engineering work.

The second schedule is the detail schedule; it is an executable schedule that is calendar driven and is used by the project management, task leaders and workers. It contains all task levels and should have measureable milestones about every one to two weeks to facilitate tracking progress. It is resource loaded, should identify critical resources and allow for contingencies, i.e. lead and lag times (float in NASA nomenclature) should be maximized as much as schedule constraints allow and the critical path should be identified.

The third schedule is a weekly task schedule prepared by each worker and used by the worker to schedule his/her work in order to meet the milestones on the detail schedule. This task schedule can be on a 3 x 5 card or any other tool that the worker finds effective. This schedule must be more than a list of tasks. Workers should schedule their time hourly for an entire week at a time and with contingencies to account for the unexpected interruptions that will happen. Many experienced engineers do this automatically but young or inexperienced engineers may need mentoring before they develop the habit of scheduling by the week.

The detail schedule is a critical document to get right because it is the primary tool used to assign budgets and workers to the project and to track the work compared to planned schedule and budget. Both the Master schedule and the detail schedule must map to the Work Breakdown Structure (WBS) that assigns tracking codes to each task for the financial management system in use. The WBS should be based on the hierarchical structure of the product and processes, i.e. the **Product Breakdown Structure** (PBS) in NASA nomenclature, or **System Breakdown Structure** (SBS) in IEEE 1220 nomenclature, so that it is tied as much as possible to physical products rather than functions. Readers are referred to section 4.3 of the NASA *Systems Engineering Handbook* for a discussion of how to structure the PBS and WBS.

The detail schedule must have all the links and precedent relationships between tasks identified. Fortunately there are software tools, e.g. Microsoft® Project, that facilitate preparing a detail schedule. Such tools are sophisticated and considerable experience is needed to utilize them properly. This often results in persons with skills in the scheduling tool being assigned to prepare or assist in the preparation of the schedules, which can lead to problems, as described next.

A big mistake inexperienced project leaders make is scheduling the project before the tasks are defined. It's ok to develop the Master schedule before defining and structuring lower level tasks that belong in the detail schedule because often Master schedules contain project constraints that must be accommodated. Inexperienced project leaders sometimes try to combine two planning steps by listing the tasks to be performed and working out a detail schedule for the list or even developing the task list on the fly using the scheduling tool. Defining tasks should be done first, at least for complex projects, because it's not obvious how tasks should be sequenced and structured until tasks are fully defined. Defining tasks means defining the inputs, the tasks that are the sources of the inputs, the outputs and the tasks the outputs support. Defining tasks is work that requires analysis by systems engineers, other specialty engineers and operations personnel. Just having a list of tasks does not mean that the inputs, outputs and the work needed to turn inputs into outputs are understood. It should be obvious that schedules constructed without such understanding are not likely to be executable without numerous corrections.

Another big mistake inexperienced project leaders often make is requiring detailed schedules for entire projects as part of the beginning planning. This is a mistake because there is insufficient information to construct such detailed schedules so that a detailed schedule becomes useless within a month or two and re-planning is required. A rolling wave scheduling process that avoids this needless rework is described after discussing tools for defining tasks.

4.3.1 Defining and Sequencing Tasks – The NASA *Systems Engineering Handbook* discusses in section 4.4 the use of **work flow diagrams** (WFD) for developing schedules. The WFD is a graphical representation of tasks, task precedencies or dependencies and "products or milestones" that occur as a result of one or more tasks. Using the WFD as defined can lead to scheduling problems due to not fully defining each task. It's possible to expand the definition of the WFD so that tasks are fully defined. However, there is a simple tool for more completely defining and sequencing tasks that can be used in conjunction with or independently of the WFD before any schedule below the top level Master Schedule is developed.

A simple and effective tool for defining and sequencing detailed tasks is the **N-Squared diagram**. This diagram was invented by Robert J. Lano and first published in an internal TRW report in 1977 according to Wikipedia [4-1]. N-Squared diagrams are traditionally used in defining software and hardware interfaces but are just as valuable for defining interfaces between hardware or tasks. N-Squared diagrams are constructed for tasks and subtasks as a nested set using a spreadsheet tool. Tasks are listed in squares along the diagonal set of squares in an N x N matrix of squares. In addition to putting the task name in a diagonal square other information useful to the project team can be added; e.g. the Work Breakdown Structure (WBS) number of the task, or perhaps the task leader's name, or a paragraph number in a proposal or task description document that describes the task.

N-Squared diagrams can be developed top down or bottoms up or using a mix of bottoms up and top down. Here it is described as a top down process but don't be constrained by this description. Start with a list of N tasks of the same level in the order you think they need to be executed. Top level tasks often have multiple levels of subtasks. List each top level task in turn in one of the squares along the diagonal of the N x N matrix. The N-Squared diagrams for the subtasks of each task are constructed as separate diagrams.

The next step is to write the external inputs to the tasks in the row above the diagram and above the square containing the task that the external inputs feed into. Then write the outputs of the first task in squares that are to the right of the first square in the first row of squares. Each output is to be in the square directly above the diagonal square containing the task that requires the output from the first task as an input. Then continue this process for each diagonal square. Outputs that feed tasks in diagonal squares that are above the row go in boxes to the left of the diagonal square and outputs that feed tasks in diagonal squares that are below the row go in boxes to the right of the diagonal square. This process defines the tasks in that the inputs and outputs of each task are identified along with the sources of the inputs and the tasks that need the outputs.

After completing this listing of outputs for each diagonal square examine the form of the resulting N x N matrix. A simple example with six tasks, external inputs to tasks 1 & 4 and two outputs from tasks 5 & 6 is shown in Figure 4-1. In this figure the tasks are labeled with numbers along the diagonal and the outputs from tasks are labels with letters. Arrows can be added to make it clear which task each output feeds, as shown in Figure 4-1.

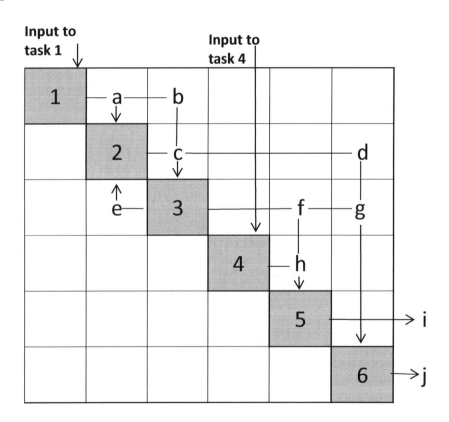

Figure 4-1 An N-Square diagram is a useful tool for defining tasks and the relationships between tasks.

An ideal sequence of tasks for a perfectly sequential project has outputs only to the right side of the diagonal. Typical projects have some tasks with outputs on the left of the diagonal squares. These outputs represent feedbacks to earlier tasks. Typically in executing the tasks in the diagonal squares that have feedbacks from tasks in squares in lower level squares assumptions or estimates are made for the inputs needed from lower level tasks and the tasks are reviewed or repeated after the lower level tasks are completed. Thus the form of the N x N diagram defines how efficiently the tasks are ordered. If there are a lot of outputs on the left side of the diagonal it is necessary to see if reordering the tasks results in moving more outputs to the right of the diagonal and thereby reducing the number of feedback loops that must be executed. Executing feedback loops increases the time and cost for completing the work. Therefore sequencing the tasks for a minimum number of feedback loops usually means achieving a task sequence that leads to the least rework, the shortest schedule and the lowest cost.

Often it is easy to get the top level summery tasks in the most efficient order even without using an N x N diagram. The value of this tool comes in sequencing lower level tasks. Once the top level tasks are sequenced then prepare N-Squared diagrams for each of the top level tasks. For example, suppose one of the top level tasks is System Trade Studies. Complex systems can have a dozen to several dozen system trades that are highly interrelated. It takes only a few hours to define the best order for the trades using an N-Squared diagram. Having just one or two tasks out of order can increase the time for completing all trades by days or even weeks.

Figure 4-2 illustrates an N-Squared diagram for the tasks defined in Figure 4-1 with task 1 sequenced to follow tasks 2 and 3. In this figure only arrows are used to indicate outputs from tasks and where the outputs are needed to make it easier to see the effects of sequencing errors. The sequence in Figure 4-2 is obviously less desirable than that of Figure 4-1 because there are multiple feedbacks, which increases the time needed to execute these six tasks compared to the time for the sequence shown in Figure 4-1.

There is a second payoff from constructing an N-Squared diagram. The form of the diagram not only defines the most efficient sequence for tasks it also shows which tasks can be conducted in parallel. For example, note that task four has externally available input data and requires no data from any of the first three tasks. Thus the fourth task can be started in parallel with the first task, or at any time such that its output is available at the start of task five. Obviously tasks can be started as soon as the necessary inputs from earlier tasks are available and staff is available to execute the task. For example, in Figure 4-1 if there was no output from task 3 feeding to task 6 then task 6 could be executed any time after task 2 and the repeat of task 2 with data from task 3 is complete. The form of the N-Square diagram reveals immediately all tasks that can be started in parallel. Starting tasks in parallel increases the flexibility in staffing. Achieving maximum schedule flexibility, i.e. maximum lead and lag times for tasks, should be one of the goals of scheduling tasks.

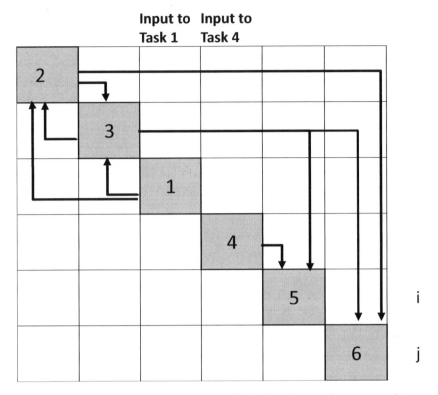

Figure 4-2 This sequence has three outputs to the left of the diagonal compared to one for the sequence shown in Figure 4-1 and therefore takes more time to execute all six tasks.

To avoid the rework resulting from too much detailed task planning and scheduling in the initial project planning effort prepare second and third level N-Squared diagrams for only those top level tasks that are likely to be executed in the first two to four months of the project. Second and third level diagrams for summary tasks that are to be executed later should be developed with sufficient lead time before these tasks are to begin to allow for planning the staffing of the upcoming tasks.

Finally, N-Squared diagrams are often a very useful visual aid to communicate to customers and managers the scope and status of project work. A skilled project manager on a satellite program posted a project N-Squared diagram where it could be easily reviewed by customers and team members. He colored tasks to show task completions so that status was communicated at a glance. This gave the customers increased confidence in the project team and project management compared to just seeing schedule charts.

4.3.2 Developing the Schedule – This discussion assumes that those responsible for developing the schedule are familiar with both the principles of network scheduling, such as **Program Evaluation and Review Technique** (PERT) or **Critical Path Method** (CPM), and are skilled in using modern scheduling software tools. This material is not intended to teach scheduling but systems engineers need to be aware of some pitfalls in scheduling complex projects. A recommendation for the inexperienced scheduler is to study a text book on scheduling projects. James B. Dilworth's Pro*duction and Operations Management*[4-2] contains an excellent discussion of scheduling. Particularly important is Dilworth's discussion of PERT using probabilistic time estimates. An alternative view of probabilistic schedules is found on the web[4-3] in a blog post by Glen B. Alleman and should be studied by anyone involved in scheduling complex projects. Alleman describes the flaws in the conventional probabilistic scheduling and defines a more reliable process using Monte Carlo Simulations. Finally read Eliyahu M. Goldratt's book *Critical Chain*[4-4] or search "Critical Chain" on the web and read about **Critical Chain Project Management** (CCPM). The *NASA Systems Engineering Handbook* has a description of a process to develop a network schedule after finishing work flow diagrams or N-Squared diagrams. This description isn't repeated here but a couple of comments are needed.

First, the NASA process calls for a bottom up approach; lower level tasks for subsystems or other work products are scheduled and then these groups of tasks are integrated in the scheduling tool to form the complete detail schedule. This can lead to difficulties and a lot of rework. The rework process is described in the NASA Handbook but much of the rework can be avoided by developing the detail schedule top down.

Typically projects have schedule constraints and developing the schedule bottom up leaves to the end the work of making the complete schedule comply with the constraints where it becomes complex and error prone. It is recommended that when a project has schedule constraints or when the project team wants to meet fixed dates for major milestones and deliveries then include these constraint dates on the Master schedule. Then develop the detail schedule top down from the Master schedule. Scheduling tools can combine these two schedules into what truly can be called an integrated master schedule (IMS).

The top down approach works best if the persons developing the schedule meet with those knowledgeable of the work needed at lower levels and work out the details necessary to comply with the constraints. This is time consuming but actually takes less time than it does to fix a complete schedule that doesn't meet constraints. Plus, the bottom up approach nearly always results in having to go back to those responsible for the lower level tasks and rework the task details. This is aggravating to workers and managers.

Second, the NASA process does not discuss the variability of task durations and probabilistic scheduling that accounts for this variability. Probabilistic scheduling is most important for accurately determining the critical path. Probabilistic scheduling is discussed in a later section on critical path analysis.

4.3.3 Rolling Wave Scheduling - Projects have schedule risk due to uncertainty in the outcomes of future design and risk management actions. Therefore it is not possible to predict long term schedule details with high confidence. If long term schedules are prepared in detail then these details become inaccurate and require rescheduling after a few months of work. The detail schedule should be developed using rolling wave planning to minimize having to rework the detailed planning whenever an unexpected event changes the work and requires re-planning of future work in order to meet the intended milestones. Rolling wave planning is a method of managing in the presence of future uncertainty due to risk. Plan the schedule in detail for two to four month periods, e.g. to the next major milestone, and with less detail beyond this period. The schedule should be updated regularly to maintain two to four months of future detailed schedule. This is necessary to identify contentions for critical resources and plan for resolving these contentions. (Recall the value of progressive design freeze discussed earlier.) Note that the NASA Handbook recommends rolling wave scheduling but suggests scheduling in detail for one year periods. A year is too long and results in unnecessary rework of schedules.

A second reason rolling wave planning is effective is that risk mitigation activities should be integrated into the master schedule. Risk management is an ongoing activity; as selected risks are mitigated others take their place. Design decisions can lead to the identification of new risks that must be mitigated. Rolling wave planning and scheduling allows the dynamic risk mitigation tasks to be integrated into the overall project schedule and budget; facilitating effective management. Thus rolling wave planning allows efficient planning in the presence of uncertainty of design results due to the risks inherent in product development.

4.3.4 An Alternative Scheduling Approach - The section on defining and sequencing tasks described how to use N-Squared diagrams to define and sequence project tasks in the optimum sequence. N-Squared diagrams are an alternative to the work flow diagrams described in the NASA *Systems Engineering Handbook*. Let's assume that detailed N-Squared diagrams, or work flow diagrams, have been developed for the first phase of a project. This means that there is a top level diagram for the entire project that defines the top level summary tasks and their interrelationships. There are more detailed diagrams for the second and third levels of those top level tasks associated with the first phase of project work, e.g. from project initiation to the first major milestone. Now the job is to convert this task data into a time phased detailed schedule for the first phase of project work.

Assuming that a modern scheduling tool, such as Microsoft® Project, is used it doesn't really matter whether the final schedule is a Gantt chart or a PERT chart because the scheduling tool can automatically generate one from the other. Thus there are several paths that can be followed from the N-Squared diagrams to the final detailed schedule for the first phase of project work. The N-Squared diagrams contain all the information necessary to go directly to either a PERT or Gantt chart. However, if the person generating the schedule isn't highly experienced there is an alternate path that is often helpful where there are many feedback loops and possibilities for starting tasks in parallel. This path inserts a second diagram between the N-Squared diagram and the PERT or Gantt schedule.

This second diagram is called a **Time Blocked Task List**. It's a simpler form of the work flow diagram or the precedence diagram defined in the NASA handbook in that it only displays sequencing of tasks. Working from the detailed level N-Squared diagrams generate rows of squares with task numbers in each square like the squares and task numbers in the N-Squared diagram. Except now tasks are not confined to the diagonal squares. The idea is to put tasks that feed from one to another in sequence on the same row or adjoining row. If the N-Squared diagrams indicate that two tasks can be started in parallel then put the second task in a row below the row containing the first task and directly below the first task. These are not rigid rules; any form is acceptable as long as each output from each task is separately feeding its appropriate destination task. What is important is positioning the tasks in columns so that the columns are in the sequence the tasks must be performed, hence the name time blocked task list. Arrows are drawn between tasks showing the feed of information from task to task.

If there is a feedback loop where data from a task is needed to replace assumptions or estimates used in earlier tasks and the earlier task must be repeated after the data is available indicate the second pass through tasks with squares following the first pass task and in the same row. Use the same task number but add a letter designator showing that this block is a second pass through the task. An example of a simple time blocked task list is shown in Figure 4-3 for the N-Squared diagram shown in Figure 4-1. This diagram explicitly shows the tasks that must be repeated due to feedback loops. The position of block 4 shows that this task can start at any time between the beginning of the project and just early enough to complete as soon as task 3a completes so that task 5 is not delayed.

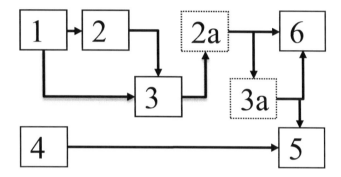

Figure 4-3 A Time Blocked Task List explicitly shows sequencing of tasks in feedback loops.

The value of the Time Blocked Task List is that it enables the inexperienced person scheduling tasks to identify more clearly when parallel tasks can be scheduled and it forces recognition that feedback loops require additional time to complete the second pass through the tasks involved. The reason that it's so important to clearly identify when parallel tasks can be scheduled is that this gives maximum schedule flexibility in scheduling critical resources. Experienced schedulers can skip the Time Blocked Task List but it's worth the effort for inexperienced schedulers to ensure that parallel tasks are clearly defined as to possible start and finish times and that time is explicitly scheduled for executing feedback loops. There is no need to develop a time blocked task list for an entire project, just block out tasks that have feedback loops and possibilities for parallel starts. Having worked out these details in the simple Time Blocked Task List it is easy to develop accurate Gantt or PERT charts. An example Gantt chart schedule for these tasks is shown in Figure 4-4.

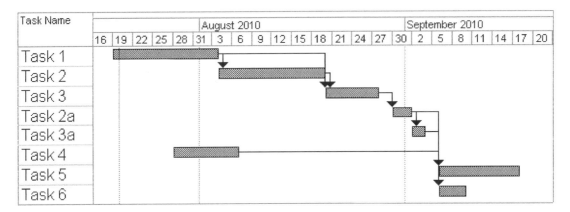

Figure 4-4 An example schedule chart for the six tasks defined and sequenced in Figure 4-1.

4.3.5 Critical Path Analysis – Critical path analysis needs further discussion because of issues that have emerged in modern complex projects. Modern software scheduling tools automatically determine and show the critical path for a project schedule. However, it is not good practice to just accept the first result for a complex project.

The simple example schedule discussed in the previous sections is a deterministic schedule; each task has a planned duration, a predecessor task and a successor task. Therefore there is a deterministic critical path. It should be obvious that the time estimate for each task has a probability associated with it. That is, there is some probability that the task will be completed in the time estimated. Therefore there is also a probability that an overall project will be completed in the time estimated by summing the times for the tasks on the critical path. Unfortunately project schedules are sometimes treated by project leaders, managers and customers as though the schedule is deterministic and any overrun in schedule is the result of poor management rather than the result of real world variation due to probabilistic time estimates. Likewise the team management is given credit for excellent management should the project come in ahead of schedule due to the same variation phenomena. Of course there is usually some truth in assigning credit or blame to the project leaders because they have numerous tools for managing a project in order to maintain schedule in spite of variations in the time to complete individual tasks. However, using probabilistic time estimates in PERT scheduling, or the alternative recommended in Glen B. Alleman's blog cited in the section on developing schedules, gives both the project leaders and those overseeing the projects a much better feel for the likelihood that the project will complete by a certain date.

Probabilistic time estimates are achieved by estimating an optimistic time, a pessimistic time and a most likely time for completing each task. The expected task duration is then defined as the mean of the distribution defined by the three estimates and is between the most likely and the pessimistic times for a beta distribution used in the standard probabilistic approach. One might expect that the critical path is the sum of tasks with zero slack in the resulting network using expected durations. But since there is a variance for each task estimate there is also a variance for the critical path schedule and therefore a probability associated with completing the project by any number of days less or more than the expected total. Since all tasks have a time to completion variance there is also probability that some path other than the most probable critical path becomes the critical path. Thus to get the most accurate probabilistic estimate of overall schedule it is necessary to analyze all paths and the probability that the project will complete by a certain date is actually the probability that all paths will complete by that date. Alternately simulations considering all paths can be carried out. A shortcut to analyzing all paths may be useful for some projects.

Consider the example of projects for developing systems with hardware and embedded software. There are often several candidates for the critical path in the early part of such projects. This is in contrast to complex projects in the past without embedded software. In the past there was typically some long lead part or assembly that drove the overall schedule so that the critical path usually turned out to be driven by the design and procurement of this long lead item. Modern digital electronics enable systems with complex data acquisition and analysis. Such systems often have special purpose digital circuits with embedded software involved in the acquisition and processing of data. The digital circuit design is usually a relatively high risk task so the designers want to build and test breadboard circuits as soon as possible so that designs can be refined and risk reduced before integrating these circuits with the overall electronics. However, these circuits require embedded software for carrying out their intended functions. Systems engineering must develop specifications for both the digital hardware and for the software. If the scheduling is done in the usual way these tasks end up driving the overall schedule in unsatisfactory ways. Managers may become confused if they expect the usual long lead parts to be driving the critical path, not risk reduction tasks.

To get estimates for scheduling the digital electronics and software tasks so that the risk reduction tasks aren't on the critical path it is necessary to get the digital designers, the software engineers and the systems engineers to sit down together and work out plans to make these tasks fit together and complete as early as possible. Even with the best efforts of the engineers the resulting schedules often end up with digital electronics, embedded software and some long lead parts all having paths with zero or near zero slack. This says that it's not possible to accurately tell which will end up the true critical path because of the variance in the time for each path is larger than the difference in slack. A way to handle such scheduling problems is to calculate the expected duration of each of these potential critical paths. It's not necessary to calculate the expected duration for all paths on such projects because it is highly likely that these few paths with near equal durations drive the critical path. Conducting the expected duration for each of these few paths also helps convince managers that the scheduling is sound even if their assumed critical path turns out to be not the most likely critical path.

The situation described for complex systems having special purpose digital electronics and embedded software seems likely to occur in other types of complex projects. If an organization is using deterministic schedules and is faced with a project schedule that has several paths with zero or near zero slack it is recommended to use probabilistic time estimates and to calculate the expected duration for the several candidate critical paths. It likely isn't necessary to expend the effort and cost to calculate the expected durations for all paths. If an organization uses probabilistic scheduling routinely then spend the effort to involve the personnel on the tasks that are candidates for the critical path so that they reach an understanding of the relationships of each other's tasks and explore approaches to workarounds that may shorten schedules as well as result in a better estimate of the true critical path.

4.4 Systems Engineering Management Plan (SEMP)

NASA's *Systems Engineering Handbook* has concise definitions for the IMP and SEMP that are summarized here. The IMP, described above, defines how the project will be managed to achieve its goals and objectives within defined programmatic constraints. The **Systems Engineering Management Plan** (SEMP) is the subordinate document that defines to all project participants how the project will be technically managed within the constraints established by the IMP. The SEMP communicates to all participants how they must respond to pre-established management practices.

Note that the SEMP relates the planned systems engineering work to pre-established practices. This means, that like the IMP, the SEMP does not have to define from scratch how the project will be technically managed; it just has to define what standard practices will be used and any deviations from standard practices. Therefore, again like the IMP, the SEMP should not be a monster sized document that no one will ever read. There is a good discussion of SEMPs in NASA's *Systems Engineering Handbook.* A concern with the NASA discussion is that it may lead teams to develop SEMPs that are much too detailed and therefore much too time consuming and expensive to prepare. The same concern exists with the discussion of the SEMP in IEEE Std. 1220. An outline for a SEMP that addresses these concerns was developed by Eric Honour of Honourcode, Inc. of Pensacola, FL in 2003.

A copy of Eric Honour's outline is presented here with his permission. The only significant change made to is to strongly recommend that we recognize that today project documentation is almost always maintained in electronic format accessible via an enterprise's intranet. Therefore one key to high quality documentation is achieving the "best" balance between copying material from one document to another and hyper linking between documents. An excess of hyper linking makes reading more tedious but excessive duplication also makes reading tedious for those that have already read other documents. A prescription for the "best" balance is not recommended here since it depends on the documentation particular to a specific organization. Authors of project planning documents should use their best judgment for achieving a good balance.

The following is Eric Honour's outline in italics with the author's comments in normal type:

The topics and outline provide comprehensive coverage. Each topic should be considered in the planning for a program, and then reviewed at the beginning of each subsequent phase. Some topics may only receive cursory consideration in early phases, with later expansion. (E.g. Producibility Analysis, System Implementation Transition)

The SEMP should contain only those processes and plans that are unique to the program.

Initial creation of a SEMP should require between 2-5 working days. Expansion of the SEMP to full planning later may take 2-4 weeks. Revision effort during later phases varies from 2 days to 3 weeks.

A complete SEMP may be as few as 5 pages, or as many as 50.

4.4.1 *Annotated Outline*

1.0 INTRODUCTION (Note that this section and section 1.1 should be identical to that in the IMP and copied from the IMP.

1.1 Overview

Identify the system and this document in one sentence. Use a sentence or two to summarize the purpose of this plan.

1.2 System Description

Describe the system, using text and/or figures as appropriate. This identical system description can be used in all plans created for this program. (E. g. define the design architecture using hardware and software trees.)

*2.0 REFERENCED DOCUMENTS (*Here is an example where the IMP should link to this List.)

List all documents that are specifically referenced within this document. Provide document number (including revision letter) and name in the format: ANSI/EIA-632 Processes for Engineering a System

3.0 TECHNICAL PROGRAM PLANNING AND CONTROL

3.1 Task Descriptions

3.1.1 Definition of Work

Define the technical work that must be performed under the contract. Show the contract work breakdown structure (CWBS), and describe the scope of each technical work package. Indicate the flow down of CDRL items to work packages (can be shown on the CWBS).

3.1.2 Subcontractor Work Effort

Define the work efforts to be performed by subcontractors or teammates under the contract. Identify the CWBS work packages containing this effort.

3.1.3 Schedule

Show the program schedule. Identify development phases and major milestones. Show a task network of task dependencies, with critical path. (Copy the one page Master schedule from the IMP and link to the detail schedule or IMS)

3.2 Organization

3.2.1 Technical Organization

Show a diagram of the technical project organization. Also include diagrams of the related program organization and functional organizations. In creating this diagram, consider the principles of concurrent engineering. (Here is another opportunity for the IMP to link to more detailed information in the SEMP.)

3.2.2 Technical Functions

Provide a textual description of the responsibilities of each person in the technical project organization. Cross-reference responsibilities to the CWBS. Include subcontractor personnel, and include subcontractor monitoring responsibilities. (Section 4.3 of NASA's System Engineering Handbook has an excellent and concise discussion of tying the lowest level of the CWBS to the responsible engineer or manager.)

3.2.3 Program Interfaces

Define the formal and informal technical interfaces to others. Consider customer and internal technical interface meetings, interface control working groups, and standards committees. Define how to control the interfaces - authority and responsibility.

3.3 Technical Control *(Don't forget the instruction in the beginning to include only things that are unique to the project. Don't describe standard processes. Just note any tailoring specific to the project.)*

3.3.1 Requirements Management

Describe the plans to manage technical issues on the program. Identify technical project meetings, issues tracking lists, issue resolution methods, and resolution authority.

3.3.2 Technical Issues Management

Describe the plans to manage technical issues on the program. Identify technical project meetings, issues tracking lists, issue resolution methods, and resolution authority.

3.3.3 Risk Management

Describe the plans to manage and mitigate technical risks on the program. Identify risk levels, threshold criteria and decision authority. Identify methods to identify risk, analyze risk and track risks. (Make the detailed Risk Management Plan a separate document and provide an overview here with links to the detailed plan. The reason is that the Risk Management Plan is a very active document that needs to be updated much more often than a SEMP as risks are mitigated and new risks are identified.)

3.3.4 Interface Control

Describe the plans to control interfaces within the system and between the system and other systems. Identify the interfaces to be controlled.

3.3.5 Configuration Control

Describe the plans to control baseline configuration of the system design. If a separate Configuration Management Plan exists, refer to it.

3.3.6 Document Control

Describe the plans to control the documents on the program. Consider master documents, copies, and distribution. Include documents created by the program team and those received from external sources. (The plans should include an "Information Architecture" diagram. Include the diagram in this section along with the descriptions recommended.)

3.4 Performance Control

3.4.1 TPM Process

Define the process to gather and calculate Technical Performance Measures (TPM). Define the frequency of measurement. Define the methods to disseminate the results, including management visibility.

3.4.2 Technical Performance Measures

Define the TPMs to be calculated. Define the raw data required and sources. Define the calculation methods for each TPM.

3.5 Program and Design Reviews

3.5.1 Review Process

Define the process to be followed for program and design reviews.

3.5.2 Review Schedule

Show a schedule for the specific program and design reviews to be used. Describe the purpose and content of each review. (The reviews should be on the Master schedule. Only describe them here if the project is deviating from the organization's standard processes for reviews. If the reviews are tailored project leaders should distribute copies of the tailored checklists ahead of preparation for each review; otherwise workers are likely to refer to standard documentation for reviews rather than to dig instructions out of the SEMP.)

4.0 SYSTEM ENGINEERING PROCESS

4.1 Process Description

In the subsections that follow, describe the program-unique processes in each phase. Use standards as a reference, and do not restate the processes therein. Consider each activity in each phase. Specifically address the automated tools that will be used and their interconnection.

4.1.1 Operational Definition

4.1.2 Requirements Definition

4.1.3 System Architecting

4.1.4 Preliminary Component Design

4.1.5 Detailed Component Design

4.1.6 Prototype Manufacture

4.1.7 System Integration

4.1.8 System Verification

4.1.9 System Validation

4.1.10 Operation and Maintenance

4.2 Related Processes

4.2.1 Electronics

Define the specific electronic engineering development processes to be used. Define the types of system engineering support required.

4.2.2 Software

Define the specific software development processes to be used. Define the types of system engineering support required.

4.2.3 Mechanical

Define the specific mechanical engineering development processes to be used. Define the types of system engineering support required.

4.2.4 Process Development

Describe the methods by which program processes will be monitored and improved. Consider the impact on systems engineering and other processes.

4.3 Trade Studies

4.3.1 Trade Study Process

Describe the process to be used in performing trade studies. Consider trade matrix formats, depth of analysis, and types of information. Describe how to handle multi-dimensional interdependencies.

4.3.2 Trade Study Tasks

Identify the major system trade studies to be performed. Indicate the issues and constraints, including interdependencies. (An N-Squared diagram can be included here or linked to. See discussion on N-Squared diagrams above.)

4.4 Requirements Allocation

Describe the unique processes to be used to allocate requirements to system components. Use standards as a reference, and do not restate the processes therein.

4.5 Design Optimization/Effectiveness

4.5.1 Analysis Methods and Tools

Describe the mathematical, simulation, and/or prototyping methods to be used to optimize the design and measure its effectiveness. Define how these methods will be used to impact the design, in which phases. Identify the specific tools to be used. Describe how to handle multi-dimensional interdependencies.

4.5.2 Design Analyses

Identify the specific analyses, simulations, and/or prototypes that will be developed. Identify when each analysis will be performed, and by which personnel. Identify the issues and constraints on each, including interdependencies. Identify the results anticipated from each. Include technical and cost analyses. Consider the following possible types of analysis:

- *Performance Analysis (Throughput, Latency, Dynamic Range, Bandwidth, Memory, etc.)*
- *Logistic Support Analysis*
- *Life Cycle Cost*
- *Design to Cost*
- *Design for Manufacturability*
- *Design for Test*
- *Producibility*

4.6 Documentation (Additional recommendations for defining documentation are given below in the section titled Information Architecture.)

4.6.1 Specification Tree

Show a hierarchical diagram of the specifications that will be created for the system and its elements. Identify the type of each specification. Indicate which specifications are internal and which must be delivered. (Replace the specification tree with a document tree or information architecture diagram for a more complete description of project documentation.)

4.6.2 Other Documents

List other documents that will be created. Indicate which documents are internal and which must be delivered.

4.6.3 Document Generation Methods

Describe the methods to be used to generate documents. Identify the specific tools, and their required interconnections. Describe the review and sign-off process.

5.0 ENGINEERING SPECIALTY INTEGRATION

5.1 Control of Engineering Specialties

Describe the integration and coordination of the various engineering specialties in each system engineering phase. Indicate how system engineering controls achieve the best mix of technical/performance values. Where the specialty programs overlap, define the responsibilities and authorities of each.

5.2 Integrated Logistics Support Plan

Summarize the approach to Integrated Logistic Support (ILS). Refer to a separate ILS Plan, if available. Consider the aspects of Logistic Support Analysis, Provisioning, and Training. Indicate the information needed by ILS and its source and time of need. Indicate the expected results and when.

5.3 Reliability/Maintainability/Availability Plan

Summarize the approach to Reliability/Maintainability/ Availability (RMA) analysis and control. Refer to a separate RMA Plan or plans, if available. Indicate the information needed by RMA and its source and time of need. Indicate the expected results and when.

5.4 Safety Plan

Summarize the approach to safety analysis and control. Refer to a separate Safety Plan, if available. Indicate the information needed by safety engineers and its source and time of need. Indicate the expected results and when.

5.5 Human Factors Engineering Plan

Summarize the approach to human factors analysis and control. Refer to a separate HFE Plan, if available. Indicate the information needed by human factors engineers and its source and time of need. Indicate the expected results and when.

5.6 Security Plan

Summarize the approach to security analysis and control. Refer to a separate Security Plan, if available. Indicate the information needed by security engineers and its source and time of need. Indicate the expected results and when.

5.7 Electromagnetic Effects Plan

Summarize the approach to electromagnetic effects analysis and control. Consider EMI, EMC, and TEMPEST as appropriate. Refer to a separate plan, if available. Indicate the information needed by electromagnetic effects engineers and its source and time of need. Indicate the expected results and when.

5.8 Value Engineering Plan

Summarize the approach to value engineering analysis and control. Refer to a separate plan, if available. Indicate the information needed by value engineers and its source and time of need. Indicate the expected results and when.

5. X (Other Plans)

(Describe other applicable specialty plans and programs such as: Test & Evaluation Master Plan; Quality Assurance; Nuclear Survivability; and Parts Control)

4.5 System Design Document (SDD)

The SDD is intended to capture an overview of the system design. It should be generated in parallel with the system design work and serve as a concise data source for new personnel coming on to a project at any time during the life cycle and as a reference document for any design modifications and future designs. The SDD should include a summary of top level requirements, but is not the source of requirements information, and it should include the project information architecture.

The SDD captures all system level design diagrams, provides a description of each state and mode, defines the operational scenarios, includes key use cases, includes a complete functional description to at least the second level and includes a complete description of the physical architecture.

If the enterprise has a mature information system that is designed to archive design data such that links are preserved then the SDD can include links to more detailed design analysis and design documentation; if not then it probably isn't worthwhile to include such links.

The SDD is not defined as part of the integrated design data packages by the DoD, NASA or IEEE handbooks. However, think of the SDD as the introduction and summary section of the engineering documentation. Just as a good technical report or proposal needs an introduction and summary an integrated design data package needs a SDD to orient users, especially on their first visits to the data package and before they begin contributing to the product design effort. The SDD is also the primary reference for future development of similar products. Trying to wade through the design data of a prior product development without having a SDD to introduce and summarize the design is a formidable task that is avoided if a SDD is available as a roadmap to the design.

4.6 Information Architecture

Design of a modern complex system results in generating a large number of specifications, documents, drawings, plans etc. A drawing tree, or a specification tree and drawing tree, may have sufficed to describe the relationships and traceability of design documentation for simple products of the past but these simple trees are no longer sufficient. It is critical to develop and maintain a data management system as part of an overall configuration and document management process so that data integrity is maintained and any desired data is easily retrievable. Part of planning a program is defining how this data base is to be configured for the system or product being developed. Unfortunately there seems to be no agreement on what name to give to the complex data base needed today. Terms like **Data Management Repository** (DMR) or **Design Database** are used for the database that includes all the documentation associated with the product development. How the database is structured and how traceability is defined is called the **information architecture.**

The program database includes subordinate databases for functions like contracts, engineering, quality, manufacturing, logistics and perhaps other specialties. Here we are concerned with the engineering database. Whether a system development is for an external or internal customer there is some documentation that specifies what is to be developed. We refer to this documentation as the source documentation. Since there can be multiple versions of the source documentation there needs to be documentation that authorizes work and identifies the version of the source documentation to be followed and any changes to the approved documentation. Here we will call this internal documentation the **Contract Database**, as it would be for an external customer.

The engineering database to be developed can be defined as three related databases as shown in Figure 4.5. Other specialty databases, such as quality, manufacturing and logistics are developed from or related to the Design or Planning Databases. Systems engineering is responsible for developing the **Requirements Database** and for a large portion of the **Planning Database**. Some appreciation for the complexity of these databases can be gained by examining how each might be broken down into constituent documentation for a system development. These examples are shown in Figures 4.6, 4.7, and 4.8.

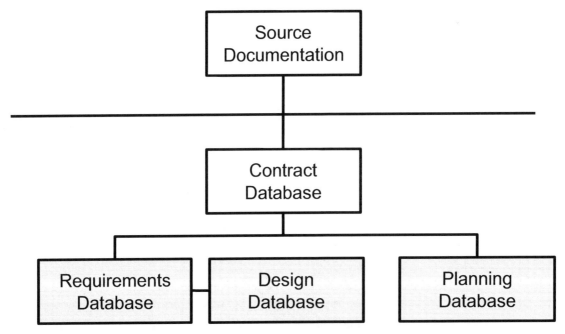

Figure 4.5 The engineering database can be defined as three separate databases for convenience.

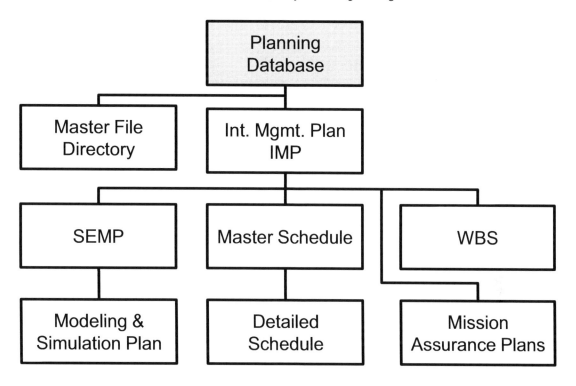

Figure 4.6 An example of typical elements in the Planning Database.

Summaries of the documentation included in the operational, functional and physical views shown in Figure 4.8 are given in section 4.3 of the DoD SEF handbook. Different systems engineering organizations may use different nomenclatures but the nomenclature used in these examples is reasonable.

There are commercially available tools for handling requirements management and for managing the entire product development data base. Examples of widely used requirement management tools include the object oriented data management tool offered by IBM® (formerly by Telelogic) called Dynamic Object Oriented Requirements System (DOORS™), CRADLE from 3SL and Vitech's CORE, popular for Model Based Systems Engineering. Requirements management tools are like wines; every experienced system engineer has his/her favorite and sound reasons for their beliefs.

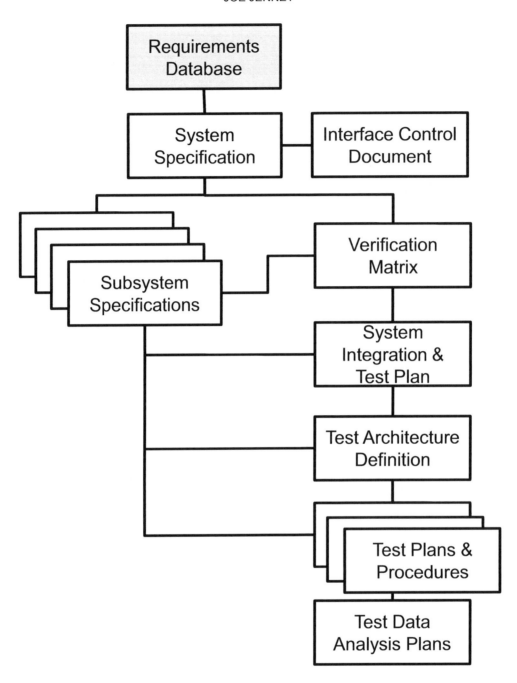

Figure 4.7 A simplified example of the constituents in a Requirements Database, not including the reports and compliance matrices associated with verification testing.

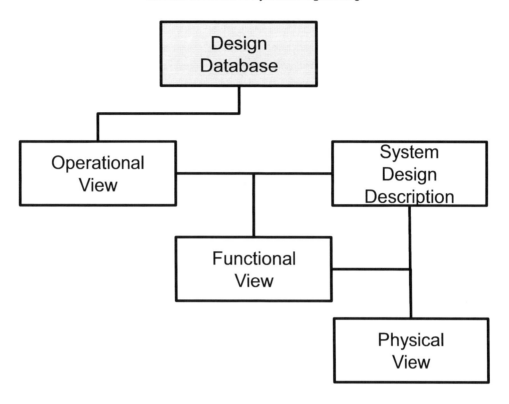

Figure 4.8 Design related systems engineering documentation can be collected in three categories, called views, along with a summary document called the System Design Description.

4.7 Modeling and Simulation Plan

Modeling and simulation should be an integral part of the development of complex systems. The DoD SEF handbook defines a model as a physical, mathematical, or logical representation of something and a simulation as the implementation of a model over time. The DoD SEF has an excellent discussion of modeling and simulation so there is no need to discuss it in detail here. But it is important to stress developing a modeling and simulation plan as part of the overall project planning process.

It can take considerable time to create or tailor models and simulations to a projects needs and to validate their performance. Thus it's important to plan the modeling and simulations as early as possible. Decisions must be made up front on the scope of intended modeling and simulation so that work won't be wasted on models or simulations that won't be used. The following list can serve to aid in identifying potentially needed models and simulations:

- System and subsystem performance analysis (typically using models initially)
- System and subsystem trade studies
- Design analyses
- System performance simulation

- Subsystem test support (hardware in loop testing)
- System integration support
- System test support
- Support of integration and test with customer systems (If system is part of a larger system)
- Support to training and/or logistics
- Trouble shooting during the systems operational life

In general, the use of validated models and simulations is highly cost effective because of improving system performance and quality and reducing testing requirements. A case in point is the aerodynamic analysis that has dramatically reduced the cost formally associated with extensive wind tunnel testing of aircraft. Also, in some cases modeling or simulation is the only way to validate performance. An example is a system that supplies measurements that are input to a software algorithm developed by a customer or another supplier and the customer specifies system performance on the basis of the output of the algorithm. If the algorithm isn't available to the system developer then modeling or simulation is the only way the customer's performance specification can be verified.

Models and simulations are so important to the development of complex systems that an enterprise's modeling and simulation capabilities have become competitive discriminators and therefore warrant investment like other potentially discriminating technologies. Some enterprises specialize in modeling and simulations that apply to many complex systems and thrive by being teammates or subcontractors on development programs. Examples include various phenomenologies like environmental backgrounds and atmospheric propagation for electromagnetic radiation that are involved in the development of modern optical and radio frequency sensors.

4.8 Supplier Relationships

Today's systems almost always involve suppliers for items varying from commodity parts to critical subsystems. Managing suppliers, particularly those supplying critical components or subsystems, is a major project activity and usually involves engineers as well as project management and sometimes enterprise management. It is sometimes ok to leave the management of non-critical commodity items to procurement personnel but critical items must involve engineers and project managers.

There are several ways to involve suppliers including strategic alliances, partnering relationships or simply contact by project or procurement personnel. Strategic alliances are for a longer term than one project; partnering relationships may be for one project or for a longer term as well. It is not the intent of this work to describe the business reasons for selecting an approach to involving suppliers but to describe why and how supplier relations should be part of project planning.

The success of modern methods like concurrent product and process development depends on involving suppliers early in the project so that the team has benefit of the supplier's capabilities in making design decisions. The objective should be to develop qualified suppliers, not just suppliers whose product data sheets offer desirable features. This means it is necessary to evaluate a supplier's technology, product quality and reliability, manufacturing and quality assurance processes and cost competitiveness. It is not unusual to find that a supplier has the desired product with the best technologies, but is lacking in key manufacturing or quality assurance processes. This can be an opportunity for a strategic alliance or partnering relationship in which process development help is offered to the supplier in return for exclusive or preferential access to the supplier's product or technology. Thus finding and qualifying key suppliers must be an ongoing process because there usually isn't sufficient time once a development program is underway. Postponing or ignoring the need to qualify suppliers ahead of time usually results in having to send engineers to the supplier to resolve a crisis caused by failure to deliver on time or with the required quality. It's much more expensive to fix supplier problems when the problems are holding up an entire project than it is to qualify the supplier before the project starts or at least before the supplier begins manufacturing the required items or implementing planned services.

Having prequalified suppliers helps reduce the number and the size of the documents related to suppliers that are commonly called **source control documents** (SCD's). An example explains this best. One manager on a schedule critical project with a dozen suppliers found that by accepting the supplier's quality assurance process the size of the SCD's was reduced from typically 50 pages to typically 10 pages; a fivefold reduction in effort for this documentation task. This manager also found that requiring the engineering team to design to their supplier's standard part specifications eliminated the need for special SCD's for these parts and resulted in a factor of six reduction in the number of SCD's needed for this example project. Thus the combination of accepting the supplier's quality assurance process and designing to standard part specifications reduced the number of pages of SCD's by a factor of 30 for this example. It may not be advisable to completely eliminate SCD's for all standard parts but engineers should consider limited SCD's for standard parts, e.g. a specification sheet plus selected critical requirements necessary to protect against changes by suppliers that could cause problems.

Assuming some type of partnering relationships are arranged with key suppliers it is necessary to recognize that these relationships require formal management. There should be documented agreements, specified points of contact and periodic reviews of the relationships. Engineers and engineering managers should regularly visit suppliers of critical components that are involved in such relationships.

It is always desirable to have multiple sources available for procured items, however, when the cost of properly qualifying a supplier and the costs of not qualifying suppliers are taken into consideration experience shows that one or at most two suppliers are all that can be afforded. Project managers must recognize that there is no shortcut. The cost of qualifying key suppliers is necessary and will be absorbed, either as preplanned and at affordable cost or as unplanned and at a much higher cost.

Finally, if possible, buy and test new critical parts before and/or during concept selection. This provides designers critical information that isn't available in a supplier's data and specification sheets. Having this data available during the design work avoids the redesign necessary when it is discovered during subsystem testing or in systems integration that the parts don't really perform exactly as the data sheets imply. This means that critical part decisions are made before design work begins or even before systems engineering is complete. The reason that this is possible is the reality that products and systems are typically designed around the use of critical parts that represent new technological capabilities or are state-of-the-art leaders. In a modern process like Integrated Product Development (IPD) the design engineers are involved from day one and can identify such critical parts at the initiation of projects.

Exercises

Consider the following questions to aid in reviewing this chapter:

1. Why do systems engineers have to participate in planning a project?
2. Why should an IMP be limited to ten pages or less if possible?
3. Why is an SDD advisable for a project that is developing a product that is typical of an enterprise's product offerings?
4. Name some advantages of using probabilistic scheduling for complex projects?
5. Why should detailed scheduling be limited to a few months ahead?

To help focus thinking about vendor relationships answer these questions for your organization:

1. Why should engineering managers visit strategic partners regularly?
2. Should internal research and development (IR&D) funds ever be invested with a supplier of critical components and why or why not?
3. When should engineers select the supplier and when should procurement select the supplier?
4. How many qualified suppliers of each component are needed?
5. How many suppliers of each critical component can your organization afford to qualify?

5 INTRODUCTION TO PATTERN BASED SYSTEMS ENGINEERING

"Each Problem that I solved became a rule which served afterwards to solve other problems" - René Descartes

5.0 Reusable Systems Engineering Products

Descartes' experience is not unique. Experienced engineers typically examine new problems to see if solutions they have used for past problems can be applied or adapted to the new problem. When past solutions can be used for new problems it both shortens the time it takes to reach a solution and increases the confidence the engineer has in the solution. Confidence is increased because the past solution has been found sound. Often the quality of a solution based on a past solution is superior to a totally new solution because the past solution has been refined with testing and use. **Pattern Based Systems Engineering** (PBSE) is a methodology of developing and exploiting past solutions for new systems engineering tasks in a standard way that allows systems engineers to reuse and share past solutions. This has an additional advantage; inexperienced engineers benefit from the work of more experienced peers and are able to work at the quality levels of more experienced engineers.

Most systems engineering work involves design elements that belong to families of design elements that design teams have experience with from executing previous development programs. The objective of Pattern Based System Engineering is to exploit this experience to minimize system engineering effort and design errors by reusing past systems engineering work much like hardware and software designs are reused. In fact, hardware and software are not reusable unless the systems engineering, particularly the requirements and decision rational behind the designs, is reusable.

Think about how experienced engineers work. When starting a new task the first thing they do is look for similarities to tasks they have previously performed. Then they refer to reports and design documentation from their previous work to guide the new work. For example, if a diagram exists from a design element of the same family as the current design element it is much faster, and results in fewer errors, to edit the diagram from the previous effort than to construct a new diagram.

Now consider if the example diagram to be edited represented not just a specific member of the family but the entire family. This means that it contains all the possible diagram items relating to the family. This complete diagram is called a pattern diagram. The editing job for a pattern diagram consists of deleting those portions of the pattern diagram that don't apply to the current design element. This can be done quickly and there is little chance of not considering any parameter since all are contained in the pattern diagram. The only possible mistakes are deleting items that shouldn't be deleted or leaving items that should be deleted; assuming that all possible items are indeed in the pattern diagram.

Note how easy it is to peer review the resulting diagram. The peer reviewers compare the pattern and edited diagrams to see if they agree with each deletion and each item not deleted. Thus having a pattern diagram dramatically shortens the time to develop a diagram for a new design effort and it reduces the chances of errors in the resulting new diagram. Experience shows that all top level diagrams can be developed from patterns in a few hours for a moderately complex system instead of the several weeks it takes to develop these diagrams starting from scratch.

Thus PBSE is largely a model based approach that produces the desired graphical models in a fraction of the time it takes to develop models from scratch or edit similar models from previous work and achieves higher quality models. Think of patterns as reusable models. However, PBSE is not entirely model based; prose documents can be part of PBSE, particularly requirements and data dictionaries containing definitions of functions. Reusability is facilitated with prose documents by structuring them as templates organized to maximize the connectivity to other documents and models and preserve traceability; thereby minimize the editing necessary for new system documents.

Readers familiar with the Yourdan systems method may see similarities to PBSE and be concerned that PBSE is subject to some of the problems of the early version of the Yourdan methods. PBSE is fundamentally different in that the patterns aren't based on a physical model and are not inherently subject to having unnecessary items in a design.

It is important to note that the title to this chapter is "Introduction to Pattern Based Systems Engineering". The objective is to introduce PBSE in a way that inspires readers to both try it and to do further reading. This book does not present systems science; rather it presents pragmatic systems engineering methods and tools the authors have found useful in their work. It isn't that we don't believe systems science is worthwhile, it's that we are working systems engineers and managers of systems engineering work and don't claim expertise in systems science. It also means that we are not trying to present a comprehensive introduction to or history of all systems engineering methods; rather we present what has proven to be effective in our work. We reference work of systems scientists and highly recommend that readers of this book dig further into methods by studying the works of well-known systems scientists.

After reading this chapter readers interested in adopting PBSE methods should study papers by William Schindel of ICTT Systems Sciences. Start with *"Pattern-Based Systems Engineering: An Extension of Model-Based SE"* [5-1] available on the web. Especially recommended is *"Requirements Statements Are Transfer Functions: An Insight from Model-Based Systems Engineering"*, by W.D. Schindel, ICTT, Inc., and System Sciences, LLC[5-2]. The implementation of PBSE presented here differs in details from Schindel's because of what we have found that works well for us and our organizations. Schindel's version of PBSE may work better than the version presented here for other organizations so we recommend readers study both. Experienced systems engineers may develop implementations that differ from both that presented here and in Schindel's work that may work better for their organizations and their work.

Note particularly Schindel's use of logical subsystem architectures for allocating requirements. Logical subsystems are analogous to the logical interfaces used to characterize inputs and outputs for systems before the physical architecture is defined. Some find this a preferable approach over traditional functional analysis and some experienced with traditional methods prefer functional analysis. For example, consider constraint requirements like mass and volume. It can be confusing for inexperienced systems engineers to allocate such constraint requirements to functions since functions have no mass or volume. Allocating mass or volume to a logical subsystem seems intuitively more appropriate.

In this work we choose to present traditional requirements allocation to functions. This approach requires less change for systems engineering organizations and in the authors' experience does not detract from the intrinsic value of developing and using patterns.

5.1 Requirements Reuse and Decision Management

Our objective for PBSE is to achieve reuse of more than diagrams, documents and models; these items are part of defining requirements and reuse of requirements is also an objective. Decisions are a central part of the requirements analysis process and reuse of decisions facilitates reusing requirements. Reuse saves time and money but it is not achieved for free. Requirements and decisions must be managed properly in order to achieve reuse. Understanding the importance of decisions in requirements analysis is helped by examining requirements analysis via the diagram in Figure 5-1. The ideas presented here are reexamined in the next chapter but this preview is provided to help explain why so much attention is paid to the details of requirements analysis in the next chapter.

It is important to recognize that the value in reusable decisions is more than facilitating the reuse of a previously used requirement or to facilitate reuse on a subsequent system development. Decisions are reusable during the life cycle of products. Problems can arise in system test or in production whose resolution requires understanding the rational for the design decisions. Potential design changes may be considered for cost savings or due to obsolete or unavailable components. In all of these cases if the rational for design decisions is not readily available and trusted then systems engineering work has to be repeated. Thus the extra cost incurred in making decisions reusable is more than recovered through savings from both reuse and from avoidance of rework.

The process of making decisions involves defining criteria, developing alternatives, conducting trades among the alternatives using a formal or informal trade study process that considers other input data to the requirements analysis process, e.g. technology readiness levels and supplier capabilities, and selecting the best alternative for the defined criteria. The NASA *Systems Engineering Handbook* has extensive discussion of decision analysis that is well worth reading. Two useful tools included in the NASA description are the decision matrix and an outline for a decision report. A **decision matrix** is a useful tool for making decisions based on weighted criteria.

Requirements from customers or market analysis drive criteria that are used by systems engineers in making decisions about system requirements and system requirements drive criteria for making decisions about lower level requirements. In describing requirements analysis verbs such as allocate, flow down, and estimate are used but all of these are fundamentally a decision making process as shown in Figure 5-1.

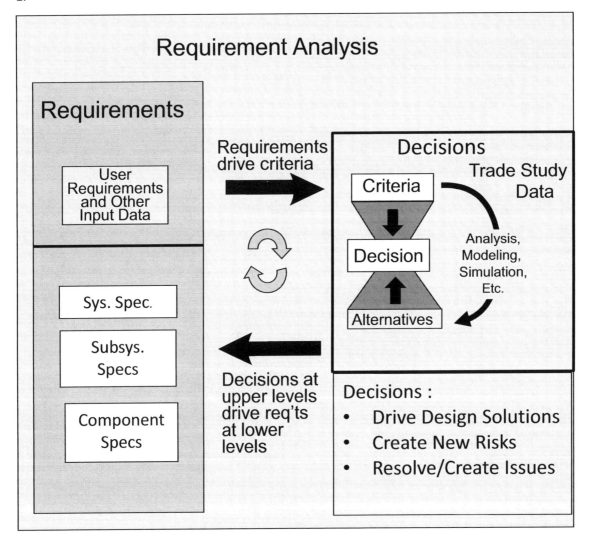

Figure 5-1 Decisions are an essential part of requirements analysis.

Figure 5-1 shows how closely requirements and decisions are connected and therefore why the desire for both reusable decisions and reusable requirements. If requirements are reusable time is saved but if the related decisions are reusable then design attributes, risks and issues driven by the decisions are understood. If the decisions are reusable then it is likely that the designs attributes are reusable; risks created by the decisions are likely to have been mitigated in previous work and issues created are likely to be known and resolved.

Achieving reusable requirements and decisions involves additional work but work that pays dividends in future developments. Figure 5-2 shows that requirements traceability is the foundation of sound requirements analysis but traceability alone is not sufficient for either sound systems engineering practice or for achieving reusability. The attributes of requirements resulting from a robust requirements analysis process that leads to control and then to reusability are indicated in the figure.

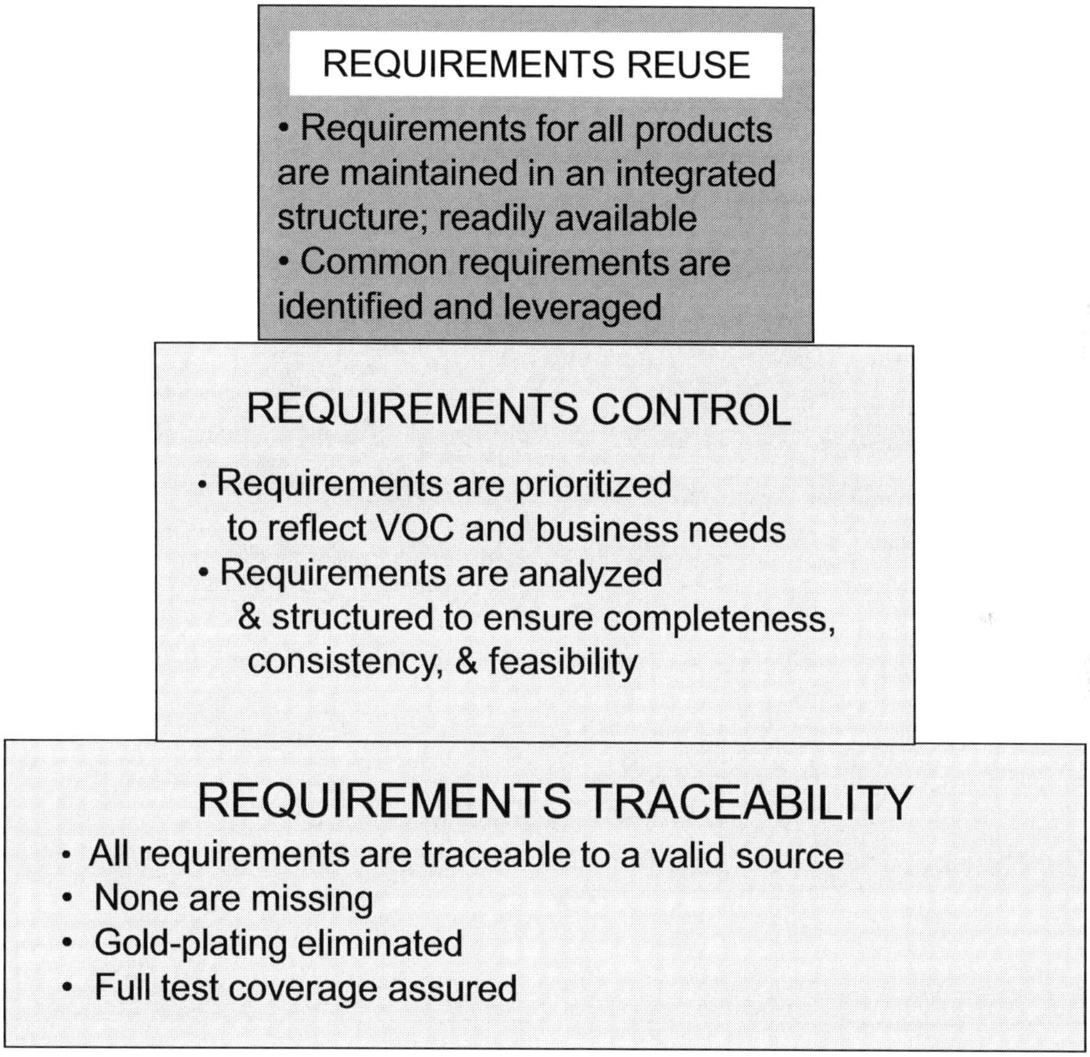

REQUIREMENTS REUSE
- Requirements for all products are maintained in an integrated structure; readily available
- Common requirements are identified and leveraged

REQUIREMENTS CONTROL
- Requirements are prioritized to reflect VOC and business needs
- Requirements are analyzed & structured to ensure completeness, consistency, & feasibility

REQUIREMENTS TRACEABILITY
- All requirements are traceable to a valid source
- None are missing
- Gold-plating eliminated
- Full test coverage assured

Figure 5-2 Achieving requirements reuse requires comprehensive requirements analysis.

Similarly achieving reuse of decisions also requires a comprehensive process as shown in Figure 5-3. The use of the pyramid format in the figures is meant to indicate that there are benefits from doing the additional work necessary to move up the pyramid. It will be helpful for the reader to review these two figures after reading the sections on requirements analysis in chapter 6.

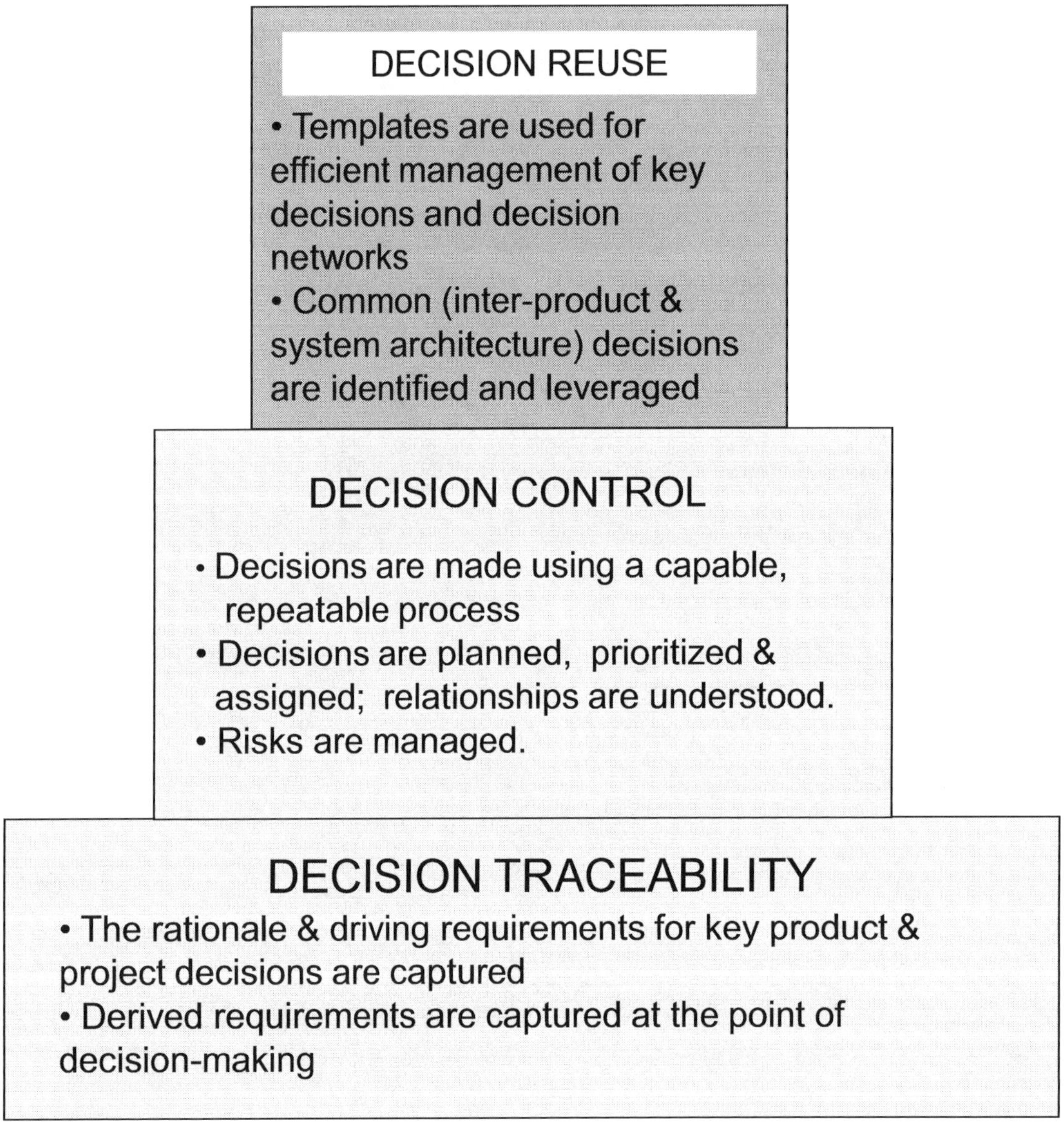

DECISION REUSE

- Templates are used for efficient management of key decisions and decision networks
- Common (inter-product & system architecture) decisions are identified and leveraged

DECISION CONTROL

- Decisions are made using a capable, repeatable process
- Decisions are planned, prioritized & assigned; relationships are understood.
- Risks are managed.

DECISION TRACEABILITY

- The rationale & driving requirements for key product & project decisions are captured
- Derived requirements are captured at the point of decision-making

Figure 5-3 Planning and analysis is necessary to achieve reuse of decisions made in requirements analysis.

The close connection between requirements management and decision management suggests the two can be integrated into a common process called data management. Data management is facilitated if the tool used for requirements management has the capability for including decision modules or links to technical memos, as do some modern requirements management tools. If simple tools like spreadsheets are used for requirements management for small systems then insert hyperlinks to technical memos that capture the decision rationale and supporting data associated with requirements. Properly capturing decision rationale and supporting data helps eliminate rework and facilitates reuse of requirements analysis work. The outline for a decision report described in the NASA *Systems Engineering Handbook* is a good guideline for properly documenting decisions and making decisions reusable.

5.2 Developing Patterns and Templates

An enterprise that sets out to develop patterns for the diagrams, documents and models used in their product lines has several possible approaches. One is to assign a team of highly experienced systems engineers to develop the patterns for a family of systems from scratch. This is an effective approach but it can be an expensive approach, even if experienced consultants are used to facilitate the process. However, it does achieve the desired patterns in a few weeks or months so that the patterns are available for near term design efforts. Often the expense is justified because having the patterns provides a competitive advantage over enterprises not using PBSE by significantly reducing the cost of the systems engineering portion of developing new products. If the experienced engineers are waiting for a new job to start then they can work on developing patterns at no additional cost to the enterprise.

A second approach is to save copies of the diagrams, documents and models from a previous design effort and treat the copies as draft patterns. For example, the diagrams for a current design effort are developed from the draft patterns in two steps. First, all new items are added to the draft pattern diagram and none of the items in the draft pattern are deleted even if they don't apply to the current design. The resulting diagrams are saved as revised draft patterns. Second, the revised draft patterns are edited to delete items that don't apply to the new design effort. These diagrams and models are saved as the working documentation for the new design effort. This approach can take considerable time to develop patterns that are complete but it has the advantage of being nearly cost free. Of course it is possible to combine the two approaches by initiating the second approach and then assigning experienced engineers that are temporarily between jobs to continuing the development of patterns. In all three approaches it is critical to thoroughly peer review the resulting patterns in order to have confidence that the patterns evolve to become complete and accurate.

Specific examples of diagrams, models and documents that should become patterns are presented in the following chapters as part of the description of the tasks systems engineers perform in carrying out the systems engineering process. This is a somewhat awkward way to present PBSE but it avoids having to present two versions of all of the data used in the systems engineering process. In some cases two or more methods are described for the same task; the intent is that the different methods can be used to verify the accuracy of work. Also individual engineers may have preferences for diagrams over matrices or vice versa. All methods should be considered candidates for patterns. Finally, in the version of PBSE presented here we include templates for text documentation as part of PBSE so that we extend the definition of PBSE beyond graphical models. We do this because many customers require text documents for items such as specifications, and because some documentation e.g. plans and data dictionaries are done in text documents. Using templates speeds the production of text documents and improves their accuracy just as using patterns speeds the production of labeled graphical models and improves their accuracy.

5.2.1 Candidate Models and Documents for Patterns – Listed here are example graphical models and documents that are candidates for developing patterns and templates that support PBSE. Some are planning documents previously describes, some are described in subsequent chapters and some are outside of the scope of systems engineering. Not all apply to every type of system development but many do and not all possible candidates are listed. It is suggested that readers consider how to develop plans discussed in chapter 4 as templates. As systems engineering models and documents are described in the following chapters consider how these items can be developed as patterns for the reader's specific organization and systems.

Pattern Diagrams for Family of Products:

- Parameter Diagrams at System and Key Subsystem Levels
- Context (or Domain) Diagram
- Logical Architecture Diagram
- Modes Diagram
- Sub Modes Diagrams
- Functional Flow Block Diagrams at Top and Second Level
- N-Squared Interface Diagram
- System Block Diagram
- Control Electronics Block Diagram (For products have digital processors and/or digital controllers)
- Other Electronics Subsystems Typically Used, E.G. Power Supplies

Document Templates for Family of Products:

- Integrated Management Plan and SEMP
- Information Architecture (Document tree)
- System Specification
- Interface Control Document
- Mapping and Transition Matrices
- Top Level Quality Control/Mission Assurance Document
- Subordinate Quality Control/Mission Assurance Plans
- System Design Description
- Typically Used Electronics Sub Systems Specifications
- Any Subsystems likely in most Products in Family
- Dictionary of Definitions

Consider what fraction of the total cost of a system development is involved in developing the models and documents listed. The cost likely varies with the type of system and organization but suppose it's ten percent. If using patterns and templates saves half the time and cost, a very conservative estimate based on the authors' experience, then the use of PBSE gives an organization a five percent cost advantage over its competition. If an organization doesn't adopt PBSE and its competitors do then the organization is likely to be at least at a five percent cost disadvantage. In today's global competitive environment cost advantages translate directly into bottom line profits or into increased market share with even greater contributions to the bottom line profits. Thus there is a high return on the investment necessary to develop and manage a pattern and template database.

In addition to the cost benefits using PBSE reduces the time needed for the systems engineering work and thereby shortens the product development schedule.

Exercises

1. Estimate for yourself the time advantages of PBSE. Time the following exercises: Construct an artificial FFBD by drawing and labeling twelve boxes using a drawing tool like Microsoft® PowerPoint or Visio. Use any set of made up verb noun combinations for the box labels. Now add arrows linking the boxes with about one half of the boxes having two arrows either as inputs or outputs and all boxes having at least one input and one output arrow. Label the arrows with any nouns, e.g. colors or numbers. Note the time it took to construct this diagram. Now time the following exercise: Starting with the diagram constructed above delete ¼ of the boxes with their labels and all their input and output arrows with their labels. Then delete about 1/3 of the remaining arrows with their labels for boxes having two or more arrow as inputs or two or more as outputs. Compare the two times. The first time is representative of the times it takes to construct a FFBD from scratch assuming one can do the thinking about as fast as it takes to draw boxes and arrows. The second time is representative of the time it takes to edit a pattern diagram.

2. Examine the tool used for requirements management in your organization to determine how it accommodates decision management or can be adapted to decision management.

3. Review the products or systems your organization develops. Group them into families that perform similar functions. Choose a family for which you expect to be developing more products or systems in the future.

4. Review the list of potential patterns and document templates presented above. Select those that your organization uses in systems engineering for new products or systems. Add any that your organization uses that are not on the list above.

5. As you read the remaining chapters select documents and diagrams from previous developments of products in the family you selected and consider how each might be further developed into a pattern or template for the selected product or system family.

6. As you read about a document or diagram spend some time developing a first draft of a template or pattern of the item for your selected family. This will help you improve your understanding of the systems engineering document or diagram and begin your organization's development of PBSE.

6 PROCESSES AND TOOLS FOR DEFINING A SYSTEM

6.0 Introduction

In Chapter 3 the product development cycle was defined in the simplest possible terms as "define, design, build, test and produce". Four systems engineering tasks were defined to be carried out over the development cycle although there is not a one to one correspondence between the five steps in the development cycle and the four systems engineering tasks. Chapter 4 discussed processes and tools for the first systems engineering task, **Plan the Program**. This chapter addresses the second systems engineering task, **Define the System**. Chapters 8 and 9 address the remaining two tasks, **Selecting the Preferred Design** and **Verifying the Technical Performance**. After some introductory material this chapter and chapters 8 and 9 cover the systems engineering process as described in Figure 3-1 of the DoD SEF handbook. Because of the four task approach being followed in this text the flow does not follow the DoD systems engineering process step by step. However, this text follows the DoD nomenclature as much as possible and describes some additional useful tools. It is hoped that the reader studies both this text and the DoD SEF handbook and that by studying the process in two different ways gains deeper insight into systems engineering.

6.1 Overview of System Development and System Engineering Processes

System development is a complex process and there are almost as many diagrams defining this process as there are handbooks of systems engineering. Before showing diagrams illustrating the **Define the System** task it is helpful to examine some of the more popular diagrams. Rather than repeat the detailed diagrams, which are available in the referenced handbooks, simplified diagrams are presented here that just show the approach to defining the development process or the systems engineering portion of the development process. The reader should always keep in mind the five fundamental steps of define, design, build, test and produce to keep oriented when examining these diagrams.

6.1.1 System Engineering Process Diagrams - Perhaps the most popular diagram historically is the **"VEE" diagram**, which is shown in the NASA *Systems Engineering Handbook SP610S* of June 1995 and attributed to Kevin Forsberg and Harold Mooz.[6-1] The VEE diagram describes the overall development process as a downward flow of decomposition and definition, connected to an upward flow of integration and verification as shown in the simplified diagram in Figure 6-1. The decomposition and definition starts at the top left with the "requirements analysis" task and ends with a tested system at the top right. This diagram is appealing because it illustrates both the "flow down" and "flow up" aspects and the sequential progress of the overall process with a rightward flow suggesting time or schedule. The time flow enables sequential tasks or design reviews along each leg of the V.

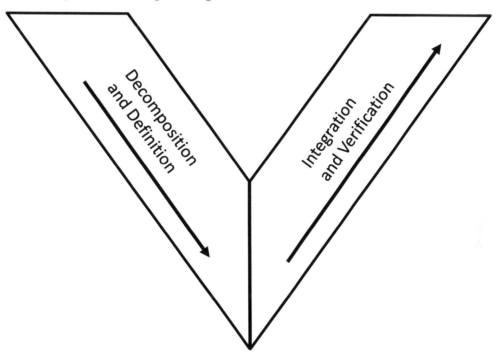

Figure 6-1 A VEE diagram illustrates tasks in two of the main activities of system development.

The 2007 version of the NASA handbook replaces the VEE with a more detailed diagram that adds the technical management and control aspects of the development process and changes some of the nomenclature. A simplified version of this diagram is shown in Figure 6-2. Although the 2007 diagram lacks some of the ascetic appeal of the VEE diagram it is more representative of the complexity of the overall process. This diagram also substitutes the more abstract descriptions "system design processes" and "product realization processes" for the more descriptive and simple terms "decomposition and definition" and "integration and verification" used in the VEE diagram. This tendency to abstract the processes for completeness at the expense of clarity is a primary reason to keep coming back to the simple and clear terms define, design, build, test and produce.

The IEEE 1220-1998 and DoD SEF handbooks focus on the system design part of the overall development process and provide more detail on this part. A simplified version of the IEEE system engineering process diagram is shown in Figure 6-3. This diagram more explicitly brings in the important trade studies and assessment activities that are necessary for achieving a good system design and lumps the technical control processes into one box labeled Control.

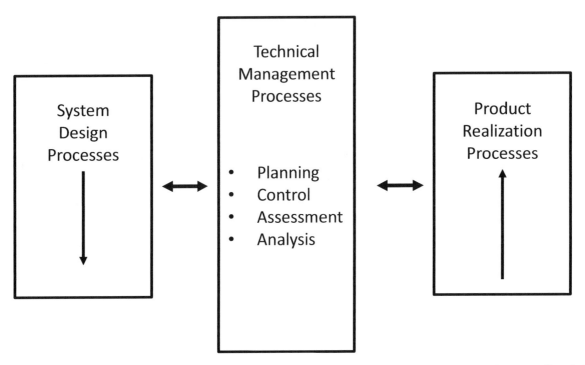

Figure 6-2 The system development process as defined in the NASA Systems Engineering Handbook SP-2007-6105 Rev 1, Dec 2007 includes the technical management processes.

Finally, the DoD handbook illustrates the system engineering process with a diagram that includes the important iterative nature of system design. A modified version of the DoD diagram that includes the technical management processes from the NASA diagram is shown in Figure 6-4. It also explicitly includes trade studies as in the IEEE diagram. Note that the diagrams in Figures 6-3 and 6-4 tend to focus on the decomposition/definition phases more than the integration/verification phases. This is a reasonable focus because mistakes or omissions in these phases are very costly but it's important to not take the integration/verification phases lightly.

The DoD diagram more explicitly represents the iterative rather than sequential nature of systems design. One aspect of design that isn't obvious in these standard diagrams is the option for design synthesis being either top down or a combination of top down and bottoms up. An example of a bottoms up design approach is designing a new car model by choosing from sets of available engines, transmissions, suspension systems and other major subsystems. Many of the requirements analysis, trade studies and assessments involved in top down design are still needed to ensure a balanced design and to verify the design but the overall development process is streamlined by having proven subsystems available for the class of system being developed. An experienced systems engineer can interpret the loops in the DoD diagram as allowing for starting a process at any point on a loop but this isn't apparent to the novice. The remaining discussion in this chapter treats the system design processes as top down for convenience but should not be interpreted as endorsing only a top down design approach.

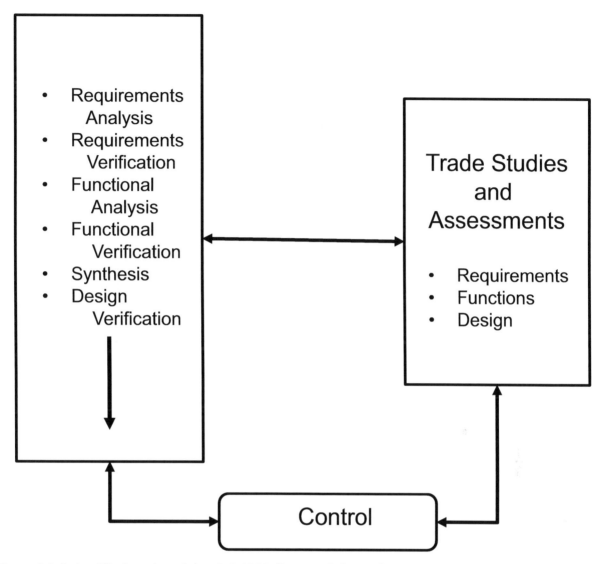

Figure 6-3 A simplified version of the IEEE 1220 diagram defining the systems engineering process illustrating three types of trade studies and analysis central to systems design.

By now the reader should have concluded that the system design process is sufficiently complicated that it isn't easily defined in any one diagram. Similarly, systems are sufficiently complicated that many diagrams are needed to adequately define them. It isn't wise to attempt to define a fixed list of diagrams and documents needed to define systems either because the complexity of a system and the maturity of the development team usually determine what diagrams and documents are necessary.

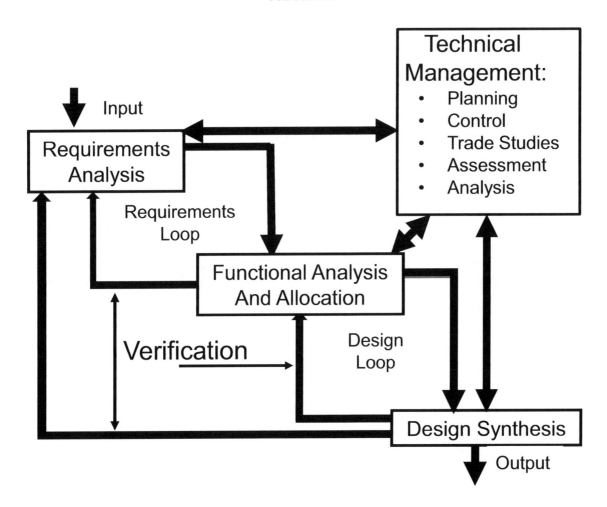

Figure 6-4 A Combination of the NASA, IEEE and DoD diagrams of the systems engineering process includes the technical management processes, the three verification processes and illustrates the iterative nature of requirements analysis, design synthesis.

The tools and processes used to define the system via the steps of requirements analysis, functional analysis, design synthesis and verification can be divided into those that treat the system as a single element, i.e. treat the relationships and interfaces between the system and its environment and other entities; and those that decompose the system into subsystems and treat the relationships and interfaces between the subsystems. (Verification is treated in a subsequent chapter as a systems engineering task called **Verifying the Technical Performance**.) Think of Figure 6-4 being the process for a system as a single entity and then think of this process being repeatedly applied to each decomposed element of the **system hierarchy**.

The INCOSE Handbook defines a **system hierarchy** as:

- System
- Element

- Subsystem
- Assembly
- Subassembly
- Component
- Part

In this material the INCOSE hierarchy is followed with one exception; the term Element is left out of the hierarchy and is used for which ever level of the hierarchy is being considered. Thus the term design element refers to the system, subsystem, assembly, subassembly, component or part being analyzed or designed.

As elements become less complex the systems engineering work for each loop becomes less and less but does not vanish. For example, the functional analysis needed for a fastener is rudimentary but it is necessary to think through, if not document, whether the fastener is required to maintain stability of the parts it is holding in three dimensions, allow translation in one or two dimensions or rotation around its axis. This is one of the reasons it is troublesome to distinguish between systems engineers and design engineers. The design engineers must execute the processes defined in Figure 6-4 to some degree depending on the complexity of the element they are designing.

Figure 6-4 implies requirements analysis comes before functional analysis. It is important to understand that this is only for the analysis of the initial requirements supplied by customers or market analysis. After analyzing these initial requirements the next step is to define the functions the system performs. For example, the NASA *Systems Engineering Handbook* identifies three steps that should precede writing requirements statements. These steps are:

- Define the design and product constraints,
- define the functional and behavioral expectations in technical terms,
- define performance requirements for each defined functional and behavioral expectation,
- then define the technical requirements in acceptable "shall" statements.

It's a lot easier to assess a set of functions to determine that the full capability of the system has been identified. Once the functions are defined and their expected performance determined then the requirements for each function are defined and documented. This also ensures that the requirements are complete and not redundant. Again, it's a lot easier to assess requirements for each function individually than to assess requirements for the whole system. Additionally, by developing requirements against individual functions, system verification is a lot cleaner. Thus the requirements loop labeled in the figure shows that requirements analysis and functional analysis are iterative.

Although a key methodology for requirements analysis, **Use Case analysis** is treated only lightly in this chapter. It is covered in more detail in a later chapter because it is particularly important for supporting model based approaches. Assessment and management of risk is such an important technical management process that it is also treated in a subsequent chapter, as is the trade study task of selecting the preferred design.

It is hoped that the complexity of treating systems engineering processes via a different set of tasks rather than the obvious choice defined by the DoD diagram shown in Figure 6-4 is offset by the benefits to the reader of having to think about systems engineering in more than one way and that this will encourage thinking that may lead to new and better methods that will be needed in the future as systems continue to become more and more complex.

6.2 Inputs and Outputs of the Systems Engineering Process

Before beginning to describe the methods and tools used to execute each block and loop in the systems engineering process as defined in Figure 6-4 it is helpful to define the inputs to the overall process and the desired outputs. A partial list of inputs is provided in Figures 3-1 and 4.2 of the DoD SEF. These inputs are partitioned into the three categories: **Inputs** to be converted to **Outputs, Controls** and **Enablers**. A more complete list of inputs in these three categories plus a fourth category of **Constraints Imposed by Internal Resources** includes:

- Inputs to be Converted to Outputs:
    - Customer Needs/Objectives/Requirements (Part of The Voice of the Customer)
    - Missions
    - Measures of Effectiveness (MOE)
    - Sys. Eng. Output Documentation (Patterns and Templates), Models, Simulations from Prior Development Efforts and from Internal Investments
- Controls:
    - Technology Base available from Strategic Partners, Teammates, Suppliers and from Internal Investments
    - Requirements Applied Through Specifications and Standards
    - Laws and Organizational Policies and Procedures
    - Customer Imposed Constraints (Budgets, Schedules, Documentation & Reporting Requirements, Other)
    - Integration & Test and Utilization Environments
    - Product/Competitive Strategies
    - Program Decision Requirements (How the customer will make decisions; including which requirements are most important to the customer.)
- Enablers:

- o Decision and Requirements Data Bases from Prior Development Efforts
- o Multi-disciplinary Product Development Teams
- o Development Organization's Domain Knowledge
- Constraints Imposed by Internal Resources
    - o Engineering and Production Skills Availability
    - o Production Process Availability and Maturity
    - o Integration, Test and Production Facilities and Capital Budget

This list does not include two items from Figure 4-2 of the DoD SEF. Maintenance Concept and Other Life-Cycle Function Planning is assumed here to be included in Customer Needs/Objectives/Requirements. System Analysis and Control is viewed here as part of the Systems Engineering Process rather than an enabler.

The list includes items contained in documentation supplied by the customer and items that are related to the organization's capabilities. The inputs are usually incomplete in that there are requirements that are necessary but not initially known. These requirements must be derived during the systems engineering work. It is not unusual for the customer's statement of needs, objectives and requirements to be incomplete, ambiguous or even contradictory when examined in detail.

As discussed in chapter 1 sometimes customers expect the development organization to complete the development of the input requirements, working closely with the customer, as part of the system development. Also, sometimes customers specify mission level requirements so that considerable systems analysis is needed to define system requirements that will satisfy the mission needs. Any constraints imposed by the enterprise's product/competitive strategies and internal resources that impact the product design should be identified and documented in the IMP during the project planning. Finally, note that some of the items on the list are easily defined, e.g. specifications and the internal constraints listed, and some are not, e.g. the organization's domain knowledge (Domain knowledge is the detailed knowledge of the customer's mission and of the systems used to satisfy this mission).

The outputs of the systems engineering process are dependent on the level of the system hierarchy and include:

- Baseline Design (System Specification, ICD, Functional Diagrams, & Physical Diagrams)
- System Design Database
    - o Requirements database
    - o System Design Document (SDD) that captures Baseline Design Diagrams and:
        - Updated Information Architecture and Document Tree

- Various diagrams, tables and matrices used to analyze, derive and verify requirements (e.g. Context Diagrams, Concept of Operation/Use Case Diagrams and QFD tables)
- TPMs, KPPs, MOEs for Baseline Design
- Design Verification Plans
- Failure Analysis Database
  - Configuration Management Database
  - Decision Database
  - Data Dictionary
- Updated Pattern &Template Database
- Expanded/Refined Models & Simulations
- Updated Risk Register
- Refined Technology Base Knowledge, including Updates to Qualified Supplier Capabilities
- Updates to Manufacturing, System Integration and Test Process/Facility Plans

Some of the outputs are new documentation, some are updates to existing documentation/databases, some are updates or new data for organizations supporting the engineering team and some is captured only in the skills and domain knowledge of the development team. Comparing the list of inputs and outputs for the systems engineering process begins to reveal the breadth and depth of work and the thousands upon thousands of decisions being made in the systems engineering work. Without the use of modern methods and tools this work can be time consuming, expensive and error prone. Shortchanging this work often causes product design problems that are far more costly than thorough systems engineering and does not build the capabilities of the enterprise for future product developments.

The system design process is fundamentally a requirements analysis process even though it is divided into the tasks shown in Figure 6-4. The output requirements are expressed in three different ways or from three different perspectives as defined in the DoD SEF. These different perspectives are called the **Operational, Functional** and **Physical views**:

- The **Operational view** describes the operational behavior of the system; how it does its job, how well and under what conditions.
- The **Functional view** describes what the system must do to achieve the required operational behavior.
- The **Physical view** describes how the system is constructed and how it interfaces with other systems and system operators.

The collection of documentation, diagrams, etc. developed under the Requirements Analysis task constitutes the Operational view. Similarly the products resulting from the Functional Analysis and Allocation task are the Functional view and the products resulting from the Design Synthesis task are the Physical view. Readers are referred to Chapter 4 of the DoD SEF for a listing of information that might typically be included in each view. Our emphasis here is on the rational for the systems engineering tasks and describing methods and tools useful in these tasks rather than trying to exhaustively list every task and every output.

6.3 Requirements Analysis

Requirements Analysis starts with the inputs defined in Section 6.2 and analyzes the mission and associated environments to establish a **Requirements Database**. Figure 6.4 shows that the Requirements Analysis Task is carried out iteratively with Functional Analysis/Allocation, with Technical Management and with Verification tasks. The NASA and IEEE documents define the subtasks of Requirements Analysis with flowcharts. The DoD SEF lists the 15 subtasks named in the IEEE flowchart and provides a description of each. Figure 6.5 is a streamlined version of the IEEE flowchart; listing the 15 tasks using most of the IEEE and DoD SEF numbering.

Several of the tasks are renamed here to explicitly include analyzing the entire life cycle for environments, scenarios and modes rather than just those associated with the system in its operational mode. Task 11 is retitled from Technical Performance Measures (TPM) to broaden the scope to include TPMs, Key Performance Parameters (KPP) and any other metrics or Measures of Effectiveness (MOE) the team or customers believe are needed. Also tasks 10 and 11 are moved to be ahead of task 12 whereas they follow task 13 in the original order.

There are five types of requirements as defined in the DoD SEF:

- Functional Requirements - The necessary task, action or activity that must be accomplished. Functional (what has to be done) requirements identified in requirements analysis will be used as the top level functions for functional analysis.

- Performance Requirements - The extent to which a mission or function must be executed; generally measured in terms of quantity, quality, coverage, timeliness or readiness.

- Design Requirements - The "build to," "code to," and "buy to" requirements for products and "how to execute" requirements for processes expressed in technical data packages and technical manuals.

- Derived Requirements - Requirements that are implied or transformed from higher-level requirement.

- Allocated Requirements - A requirement that is established by dividing or otherwise allocating a high-level requirement into multiple lower-level requirements.

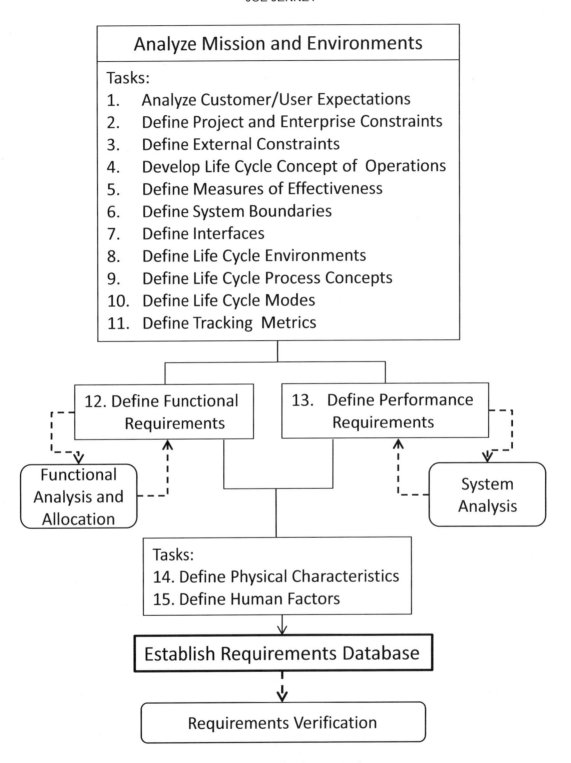

Figure 6-5 Requirements Analysis consists of 15 subtasks that analyze the mission and environment to establish a Requirements Database.

The 15 tasks are structured so that all necessary requirements of each type are identified and documented in the **Requirements Database**, which includes the Operational, Functional and Physical views. Each requirement must be traceable back to the input documentation or to a mission need derived from analyzing mission needs to ensure that no extraneous requirements are added.

The requirements analysis process is presented as a sequential process. These steps typically overlap and work previously completed may need to be adjusted as the detailed knowledge of all the requirements becomes more fully understood. Tasks are grouped in Figure 6-5 because the tasks in a group don't need to be done in any specific order and some can be done in parallel. Tasks 1 to 11 and 14 to 15 analyze the system as a single entity. Tasks 12 and 13 begin the decomposition of the system into logical functional components (task 12) and begin to allocate performance parameters to these components (task 13). Tasks 12 and 13 feed directly into the Functional Analysis/Allocation and Systems Analysis tasks and receive information back from these associated tasks as indicated in the figure.

Executing the 15 tasks involves making many decisions and defining many relationships. As the many decisions are made and relationships are examined, it is critical that they are documented and tracked to facilitate updating tasks as new information is available.

Attempting to do requirements analysis without the use of supporting methods and diagrams is like trying to do math without equations and special symbols. It is possible but much more difficult and leads to more mistakes. The methods and diagrams selected for discussion in the following sections are not meant to be the only possible choices or a complete set; rather those presented are selected because they illustrate an important part of systems thinking or offer alternatives to conventional approaches covered in standard handbooks.

A principle behind offering alternatives is that when there are two ways to do a systems engineering task or two ways to present the results it is often beneficial to use both ways. This is because one offers a self-check on the other and some reviewers are able to critique one way better than another when the work is presented to peer reviewers. Once something is complete in one format or by using one tool repeating it in another format or with another tool takes little more time than carefully reviewing the accuracy of the first way and is much better at catching errors.

6.3.1 Reflections on Requirements Analysis- Understanding the requirements of a system is an extremely important, but often overlooked aspect of systems engineering. The nature of engineers is they like to design. Engineers often think 'I know what to build', and move right into the exciting part of developing a product. Additionally, programs very rarely have the luxury of a relaxed schedule, so it is often felt for the project to be delivered on time, the design activity must start right away. Both engineers and managers are motivated to minimize the requirements phase of the program. This is reinforced by the perception that requirements appear straight forward. However, when examined from a systems perspective, requirements are not so simple and impact every aspect of a program. These impacts must be understood to eliminate rework and waste in the subsequent phases of the program.

A critical characteristic of a good systems engineer is always maintaining an integrated view of every task. Systems engineers must ensure they always look at the whole picture and assess impacts from all aspects of the program. Whereas engineering specialists focus their attention primarily on their individual domain, systems engineering must look across all engineering domains, as well as across the program elements for cost and schedule. When developing requirements, the system engineer must apply this integrated view to understand the impact of each requirement on both the design and the program. This understanding is essential to ensuring that the individual engineering specialists fully understand what drives their design

System level requirements often have unforeseen impacts on elements of the design. The systems engineer having the 'system engineering view' must be able to anticipate these impacts from the high level requirements before the design is initiated. This anticipation is important to program planning to make sure the work is accurately defined and yields a viable program cost and schedule. It is essential that requirements are balanced in difficulty. This means for example that the imaging performance of a digital camera is balanced between the lens and the detector arrays. If the requirement on the detector arrays is relaxed then the lens may be complex and expensive. Conversely, if the requirements on the lens are relaxed then the requirements on the detector arrays may require inordinately expensive detector arrays.

In the remainder of this chapter numerous diagrams are defined that aid in developing and communicating an understanding of the system being developed. Engineers unfamiliar with systems engineering may ask why so many diagrams are needed in the process of requirements analysis. One way of understanding why is to imagine that no diagrams were developed; rather written requirements were provided to designers and the designers produced a final drawing package. A moderately complex system could have thousands of drawings. Imagine trying to explain to customers that a system build to the drawings would do what the customer intended. Similar difficulties would result in trying to communicate with suppliers and managers. Human cognitive abilities just don't allow achieving an understanding of a system from large written descriptions or a detailed drawing package with thousands of drawings. Higher levels of abstraction are needed to provide understanding without overwhelming detail. Systems engineers use diagrams at various levels of abstraction to provide the understanding and the communication tools needed in system development.

All of the diagrams and models of the Operational, Functional and Physical views can be called system models because each offers some type of model of the system or of part of the system. Thus systems engineers develop system models from different perspectives and at various levels of abstraction or detail to provide the desired understanding of the system. Cognitive scientists tell us that having an understanding of a system means we have a mental model of it that allows us to interpret and make predictions about the system. The many different system models are necessary to enable us to focus on various aspects of the system without being distracted by the detail of the entire system; that is to form various mental models that are fundamental to our understanding.

Different people will form different mental models of a system. The customer's mental models are likely quite different from those of the manufacturing engineer. The software engineer's mental models are different from the hardware designer's. These different mental models are due both to human differences and to people forming mental models appropriate to problems important to them. People interpret models in a way that is correlated with their learning styles. This is one reason it is valuable to develop system models that illustrate the same information in different diagrams or formats.

6.3.2 Diagrams and Methods for Requirements Analysis - This section presents an integrated view of requirements analysis rather than discuss each of the 15 tasks, which are defined in the DoD SEF. Effective tools for analyzing the mission and environments include Context or Domain diagrams and Concept of Operations (Con-Ops) description documents. These tools analyze a system as a single entity and are useful in addressing IEEE tasks 4, 6, 7 and 8.

6.3.2.1 Context or Domain Diagrams - The first step in requirement analysis is to understand the context in which the system will be deployed. The context or domain defines the external world that interacts with the system. The entities that the system interacts with include, people, other systems, and environments. There is overlap between these entities but considering them separately assists in making sure all interactions are addressed.

The people that interact with the system include all users, operators, and maintenance individuals. Keep in mind that users may be different than operators and have different concerns. An example is the passengers on an airplane have requirements that need to be met but they are very different from the requirements of the pilot. Likewise, the team that maintains the airplane requires very different interactions with the airplane than either the passengers or the pilot. Others that need to be considered are those that install and dispose of the system. When designing a nuclear power plant, understanding how the facility materials are processed at end of life is as important to the design as how the reactor is controlled when it is operational. Lastly, during development, the designers often need access to detailed performance measures to verify correct operation of the system that is not required by anyone else.

Understanding interactions with the external world include physical elements providing input or receiving output from the system. These typically fall into mechanical or electrical areas but many products include other areas. Cameras have an optical interface to the external word. Geiger counters obviously interact with atomic particles and gamma rays via a radiation interface. Most of the interactions are readily apparent, but the context needs to include all interactions ensuring there is no undesired impact on performance. For example, many systems receive power from both batteries and the electric power grid. It is not sufficient to identify the input of power; the context must identify both sources.

Many environments are applicable to nearly all hardware systems. These include temperature, shock, vibration, acceleration, and humidity. When assessing these environments, the operational conditions must be assessed. It is typical that the storage conditions for a product may be different than the operating conditions. There may also be operational scenarios that vary for the product over the life of the product. A satellite for example needs to survive the launch environment for only a very short period of the overall life of the system. Beyond the typical environments, some systems have unique environments such as radiation, biological or chemical exposure, or electromagnetic interference (EMI).

A standard method for capturing the context of the system is with a context or domain diagram. Here the term context diagram is used. A context diagram places the system at the center, surrounded with all external entities identified by assessing interactions with human, other system, and environments. Each entity is identified separately with the type of interaction between the system and the external entity identified. The interaction identifies what flows as well as the direction of the flow. Interactions between external entities are not necessary to include on the diagram but may be included if they provide clarification. The system is identified as a single entity meaning there is no visibility as to its internal components. An example context diagram for a digital camera is shown in Figure 6-6.

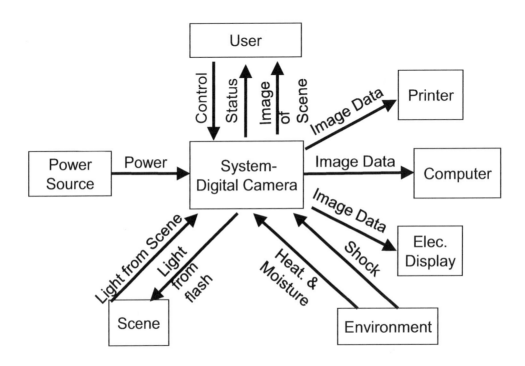

Figure 6-6 A context diagram illustrates the external entities that the design element communicates with and what information is communicated between the design element and each external entity.

A second version of a context diagram groups the information communicated between the system and the external entities according to the type of logical interface. The interfaces are logical at this point rather than physical because the physical interfaces are yet to be defined. An example of a partially complete context diagram with logical interfaces is shown in Figure 6-7. In the example shown in Figure 6-7 a sensor for some type of radiance is hosted on a platform that could be a ground, air or sea platform. If this diagram included labeled arrows for all possible information exchanges between a family of radiance sensors and any host platform then the diagram would be a pattern diagram for the family of radiance sensors.

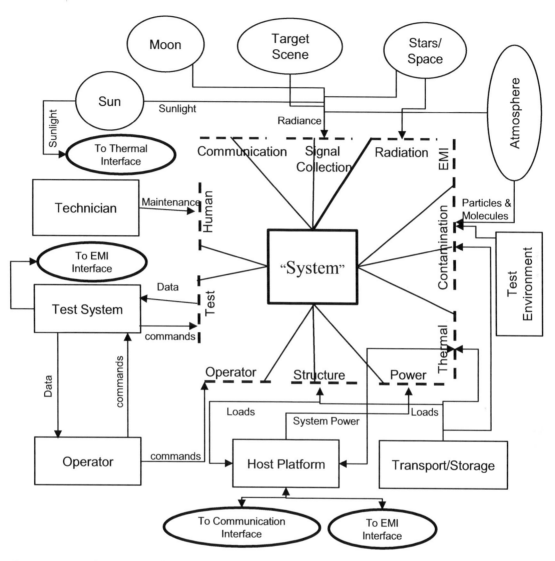

Figure 6-7 An example context diagram for a sensor system mounted on a host platform illustrating the logical interfaces the system has with the external entities.

Having a complete pattern diagram at hand the systems engineers for a specific sensor on a specific host platform can quickly develop the context diagram for their system by examining each arrow and deciding if it belongs or should be deleted. Deleting the arrows not required is much faster than adding arrows to build a diagram and the risk of missing any pertinent information exchange is lower. In constructing a context diagram with logical interfaces it is helpful to color code each interface and the arrows belonging to each interface. The example in Figure 6-7 is not meant to be a complete pattern diagram, but rather to illustrate the approach to constructing a complete pattern for a context diagram representing a family of systems.

Interface control drawings or interface control specification documents (the acronym ICD is used here for either the drawing or the document) are an essential part of system design. The ICD captures the quantitative details of the information exchanged. Comparing the context diagram against the ICD is a quick way to check that every arrow to a logical interface on the diagram is accounted for in a specification in the ICD.

Context diagrams help define the design task by confirming the boundaries of the design problem, identifying the logical external interfaces and the information exchanged with all environments and entities external to the system throughout its life cycle; thus laying the ground work for defining functional requirements and developing the ICD.

The information included in a context diagram can also be presented in a matrix format as shown in Figure 6-8. The matrix format has the advantage of being less cluttered and therefore easier to review but it does not include the direction of information without the extra work of adding tiny arrows, the words "from" and/or "to" or color coding each cell.

In constructing a context diagram or matrix it isn't necessary to be quantitative in defining the information exchanged between the system and external entities. The idea is to ensure that all of the information and external entities are captured. It's ok to use general terms like "commands", "data" and "loads" if the diagram is highly cluttered. Specific details on such information are to be provided in the ICD, but if it doesn't make the diagram too cluttered to be useful add qualitative detail. For example differentiate between test commands and operating commands and define the sources and types of loads. Some may find it useful to add more specific information in the information cells of a context matrix or even to add links to cells as reminders for use later in developing the ICD. Some may find it useful to construct both a context diagram and a context matrix as cross checks to ensure complete capture of information exchanged with all environments and all external entities.

Just as for the diagram format if a complete matrix for a family of systems is developed as a pattern then the matrix for any specific system is quickly developed by deleting information exchanges not pertinent from the pattern matrix. A matrix for a specific system can be developed from a pattern matrix in a fraction of the time it takes to generate a new matrix from scratch and is more likely to be accurate.

Following completion of the context diagram, a diagram data dictionary is generated or updated. This dictionary defines all elements of the diagram, as well as all interactions. The dictionary should include as much understanding of the element as possible. The reader of the dictionary needs to be able to understand what is included or not included in each element.

External Entities	Interfaces									
	Operator	Structure	Power	Thermal	Test	Human	Signal Collection	EMI	Contamination	Communications	Radiation
Host Platform		Loads	System Power	Heat				EMI	Particles & Molecules	Commands, Data	
Transport/Storage		Loads		Heat					Particles & Molecules		
Operator	Commands										
Test System	Commands, Data		System Power		Commands, Data			EMI		Commands, Data	
Technician						Maintenance					
Test Environment				Heat			Radiance	EMI	Particles & Molecules		
Target Scene							Radiance				
Atmosphere							Radiance		Particles & Molecules		
Sun				Heat			Radiance				
Moon							Radiance				
Stars/Space				Heat			Radiance				Charged Particles

Figure 6-8 The information in a context diagram can also be presented in a matrix of interfaces vs. external entities.

6.3.2.2 Life Cycle Concept of Operations - The context diagram defines the system interactions with the external world. However, this is a static view. A dynamic view of the system is also necessary to understand how the system interacts with the external world. This dynamic view is provided with the **concept of operations** (Con-Ops) description. Although the emphasis is on scenarios describing normal operation of the system sometimes scenarios describing operations in unique conditions or non-operational conditions should be included. Examples might be if the system has unique operations during testing, maintenance or at end of life that add to the system requirements.

The concept of operations includes a number of **scenarios** that in total convey an overall understanding of the system behavior. It does not include an exhaustive list of every operating mode of the system, but provides a broad view that the systems engineer uses as the basis for defining the details necessary in the design requirements. Each scenario provides a single thread of sequential events that starts at a typical predefined state, and completes when a primary set of operations is complete. Scenarios are success oriented, providing a linear flow for operation of the processes of interest, and do not include alternate processing in the flow. Following the completion of the scenario, faults can be described as an alternate flow describing how the system may react should a fault occur. During the scenario, the flow may describe interactions with the operator. The flow assumes a particular input from the user thus providing the singular linear flow. Should alternate inputs provide important understanding of the system, they need to be described in separate scenarios. If there are many variations of processes that all start with a common flow then it may be convenient to describe the initial common portion as one scenario, and the multiple variations as independent flows.

Scenarios need to include assumptions about what may have occurred prior to the start of the description, and describe the state of the system at the completion of the description. Environmental conditions need to be defined if they affect the operation of the system.

Use cases, described in a later chapter, are an excellent standard tool used to develop scenarios.

Typical operating scenarios include:

Initialization – The initialization flow starts with the system unpowered and completes when the system is ready to perform its primary function. There may be multiple initialization scenarios. For example, when a computer boots, the computer may prompt the operator for input as to whether to perform a standard boot, or to perform a maintenance 'safe' boot. These two scenarios are described separately, both starting the same way with the application of power, but deviating based on the input from the operator.

Operation – Starts at the completion of initialization and completes when a process of the system is complete. The process may be scheduled, initiated by an operator, or the result of some other event detected by the system. An operational scenario may include a number of processes but should be short enough to be easily grasped by someone not familiar with the system. Multiple operational scenarios may be necessary in order to cover all primary processes.

Maintenance – Starts in a normal operating state and completes when the system is returned to the operating state. Assumptions of events leading up to the maintenance action are a particularly important part of these scenarios.

Repair - Starts in a fault condition and completes when the system is returned to the operating state. It is not possible to include scenarios which address every possible fault. The key is to remember that the intent is to provide the systems engineer with the broad understanding of how 'typical' faults are handled. This should provide enough understanding to extrapolate the operation of faults not explicitly described.

The content of the concept of operations must be developed with support of the user when possible. This document captures the user expectations for how the system behaves. Additionally, this is the opportunity to guide the user to alternate operation if the behavior can be simplified. In the end, a user review is critical to receiving formal agreement with the final document. This document forms the basis for the system requirements for behavior. Additionally, it is the primary input to any user's manual.

6.3.3 Define Functional Requirements - The context diagram and the concept of operations provide an understanding of the static and dynamic interactions of the system with the external world. Identifying **functional requirements** requires that we begin to examine what the system does. This first involves looking at the system as a complete entity and then decomposing top level functions to lower level functions. It includes examining the **modes** that a system exists in throughout its life cycle. These analyzes set the stage for allocating functions to physical elements; the work of the design loop. A useful tool for examining function for the system as a single entity is the Parameter Diagram.

6.3.3.1 Parameter Diagrams - The **Parameter Diagram** (P-diagram) supports requirements analysis by determining and communicating the function of the system or any design element, defining those parameters that can cause undesired performance (**noise factors**) and defining the **control factors** the element has (or should have) whose values are selected to achieve the desired performance of the ideal function and to minimize the sensitivity to the noise factors. P-diagrams are useful at any level of product design, from mission level to part level. The form of a P-diagram is shown in Figure 6-9.

Defining the ideal input and the control factors is usually straight forward. The real work comes in defining the ideal response of a design element, the undesired responses and the noise factors that cause the undesired response(s). The design element, labeled "system" in the figure, is treated as a single entity and the control factors selected as appropriate for the level of the design element. Thus, if the design element is at a mission level the control factors relate to the various systems that interact to fulfill the mission. It isn't helpful to list screw thread size as a control factor for a mission or complex system design element but screw thread size may be a control factor for a design element at the component level.

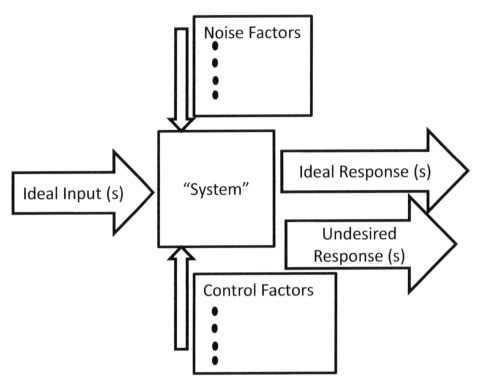

Figure 6-9 A P-diagram defines the ideal function of a design element and lists the noise factors and control factors that must be controlled in the design to achieve as close as practical the ideal function.

Consider the P-diagram for a pencil. The ideal response to the input of hand motion is a line on a writing surface of a uniform density and width. However, a pencil has another ideal response that is important to users. It should sharpen to a fine, uniform point in response to being sharpened. Undesired responses include skipping, tearing paper and the point breaking during sharpening. Control factors include parameters like graphite composition, hardness and diameter and the type of wood used for the casing. Noise factors include writing surface roughness, writing surface material, type of pencil sharpener, sharpness/dullness of pencil sharpener blade(s), temperature, humidity and aging. These lists aren't meant to be complete but to illustrate the principles involved. Note that variations in composition or trace impurities are not listed as noise factors since the purity and composition are controllable. Sometimes it is difficult to decide whether a parameter is a noise or control factor. It's best not to waste time trying to be to pure with definitions; capture the parameter as one or the other and ensure it is taken into account during the design.

Generating the P-diagram is a simple and convenient method for the engineer, whether a system engineer or a design engineer, to think about the effects of the environments the design element faces in its life of use, the types and influence of other elements it interacts with in test, use and storage; the ways in which the element is used (intended and otherwise) and the expectations of users. Thus the P-diagram complements the context diagram and the concept of operations and thereby supports requirements analysis as well as functional analysis.

The reason for listing the undesired responses of a design element is that the preferred method of controlling undesired responses is to optimize the ideal response in the presence of the noise factors. This is the essence of robust design practice discussed in the books of Clausing and Phadke previously cited. Design practices prior to robust design often attempted to reduce the undesirable responses directly. Directly minimizing one undesired response can result in redirecting the energy causing that undesirable response so that it causes another undesired response. In contrast robust design practices maximize the ideal response in the presence of the noise factors so that all the undesired responses are controlled.

One of the benefits of developing a P-diagram is that it defines the parameters to be considered in robust design. Another is its communication value. The P-diagram communicates essential parts of the design problem to both the design team and to any reviewers of the design work. If the design element is of the same family as something that the team has designed before then it takes little time to edit the pattern P-diagram from a previous design by adding any new factors and deleting any factors that don't apply to the new design element. It is likely that the ideal inputs, ideal responses and undesired responses are very similar and take only minutes to edit. However, it is the differences between the new design element and those the team has designed before that are important to define at the outset. The P-diagram is a quick way to define and capture those differences.

If the design element is entirely new to the team then the P-diagram is critical to identifying all of the noise factors that must be accounted for in the design. Often the effects of noise factors on the performance of a design element are analyzed individually or in groups. Color coding the noise factors on a P-diagram is a quick way to remind reviewers of which noise factors are under consideration in a particular analysis, which ones have been previously evaluated and which ones remain to be analyzed. Using robust design practices such as Taguchi design of experiments noise factors can be grouped or compounded to speed the analysis. The P-diagram is again a good communication tool to describe the structure of a Taguchi analysis of variance experiment. (Taguchi design of experiments is described in detail in the book by Phadke cited earlier.)

Examining the interactions on the context diagram leads to the identification of the functions performed by the system in addition to the ideal function and undesired interactions defined in the P-diagram. Each interaction on the context diagram is reviewed to define the inputs and outputs from the system. The transformations performed on the inputs, and driving the outputs, establish the functions performed by the system. Functions should be defined following the classic engineering view that outputs are defined as a transform on one or more inputs. When defining system functions, the requirements must define every output that is produced from every combination of the legal set of inputs. The definition of the function transformations must be comprehensive and accurate. The functions must provide exhaustive definitions. Requirements often are not exhaustive by not covering all possible outputs that can be produced.

The concept of operations (Con-Ops) does not provide an exhaustive understanding of all operations of the system, so the identification of all functions must rely on the context and P-diagrams. Additionally, interactions on the context diagram may be involved in multiple functions. Involvement of the user in the function definition provides confidence that all system functions are identified. Con-Ops and P-Diagrams typically focus on the functions which produce the target outputs which the user desires from the system. However, there are often other functions that must be performed to support the operation of the system but that don't directly produce the target outputs desired by the user. For example, data systems typically employ error correction algorithms to ensure the desired data performance is achieved. Users typically do not request an error correction function as part of the system. The system engineer must provide the broad vision, based on his understanding of the user needs, to recognize this is a necessary function which the system must perform in order to ensure the algorithm performance requirements are specified. This highlights an important value of the context diagram. Examining all interactions of the system with the external world usually identifies functions the system must perform that are not typically identified as requirements by the user.

When examining the interactions on the context diagram, it is also important to remember that some of the interactions will map to a real physical interface. Data flowing in and out of the system usually represents a physical interface. However, some interactions, such as radiated EMI, are not mapped to a physical interface. It would be a lot easier to design a system if the radiated EMI interactions could be allocated to a specific connector but in this case, the functions need to be specified with the understanding that many mechanical characteristics have requirements that address this function. Functions that do not deal with explicit physical interfaces are often difficult to specify.

The detailed understanding of system interactions is translated into the final functional requirements. For each function, each individual parameter for performance, behavior, and visibility becomes an independent, unique requirement. Having identified functions, detailed understanding ultimately leads to functional requirements. The specifics of the performance, behavior, and visibility associated with the input, transform, and output must be included. Performance aspects include tolerances, uncertainties, response times, bandwidth, data structures, etc. Behavior includes considerations of how functions are affected by the operating state of the system, other functional results, and environmental conditions. Visibility defines how much knowledge is known about how the function is performed. For example, an error correction function may be required by the system. This error correction function may be defined simply by the ability to perform a specified level of error correction, or the system may need to implement a specific algorithm.

When all functions are defined, the customer must review and agree on the result at a System Requirements Review (SRR) or System Functional Review (SFR). Other tools and methods useful in identifying functional requirements include Quality Function Deployment (QFD), defining system modes and Kano diagrams. These are described in the following sections and a later chapter.

6.3.3.2 Define Life Cycle System Modes - Most texts, e.g. the NASA SE Handbook, use the terms **states** and **modes**. These terms are often confused because both have transition diagrams and are often used interchangeably. It is preferable that they are not the same thing and are not interchangeable. Both state and mode diagrams define a condition of the system identified as a box, with a transition between them. In both cases, the transition is labeled with a system event which triggers the transition. The difference between a state and a mode is the definition of the box. States define an exact operating condition of a system, whereas modes define the set of capabilities or functions which are valid for the current operating condition. Let's use a simple example of a diagram for a television with two conditions OFF and ON. There are two transitions between them, one pointing to ON labeled as remote control ON detected, and a second pointing to OFF labeled remote control OFF detected. This diagram can represent a State or a Mode diagram. So what's the difference? If it is a Mode diagram, there is a data dictionary defining what functions can be supported in each condition. The TV in OFF mode must perform the function to sense the IR signal from the remote control to turn on. The TV in the ON mode must perform functions to change channels, adjust volume, and sense the IR signal from the remote control to return the TV to off. Knowing the TV is in the ON mode does not provide knowledge of what exact function the TV is performing. If this is a state diagram, there is no understanding of what functions the TV can perform. The state diagram details a single flip flop that toggles state based on the inputs of remote control ON, or remote control OFF. The output of this flip flop probably is used to activate primary power. In a more complex state diagram, there are many more memory elements but the key point is, in a given state, the condition of every memory element is explicitly known.

System modes have a hierarchy similar to functions. Modes can be decomposed into sub modes, which are also called modes, just as functions can be decomposed into lower level functions. Modes and the transitions between modes can be shown in a hierarchy of diagrams and matrices. The top level diagram or matrix should contain all modes in the life cycle of the system from assembly to an end of life mode. An example of a simple life cycle mode diagram for a commercial product for home use is shown in Figure 6-10. In this figure the event that determines transitioning from one top level mode to another is included as a labeled arrow on the diagram. It's important to define and examine the modes for all of a systems life cycle rather than just its IN USE mode. Sometimes there are requirements that are necessary for integration and test, storage or end mode that are not included in the modes for the IN USE mode and the life cycle mode diagram is useful in identifying such requirements.

Diagrams or matrices can be developed for each of the top level modes that have sub modes to define the functions required in each mode and the allowed transitions between functions. Thus the top level mode diagrams identify lower level modes that are then examined during functional analysis to assist in defining all of the functions required in every mode of the life cycle of the system. This topic is revisited after discussion of functions and functional decomposition.

Transitions between sub modes are usually more complicated than between the top level life cycle modes. A very simple example that illustrates this is shown in Figure 6-11 where the transitions among the three modes of the IN USE mode of the product having the life cycle modes shown in Figure 6-10. The transitions are not labeled in this diagram at this point because the events causing these transitions can be dependent upon the physical design. It is only necessary at this point to define the allowed and required transitions between modes.

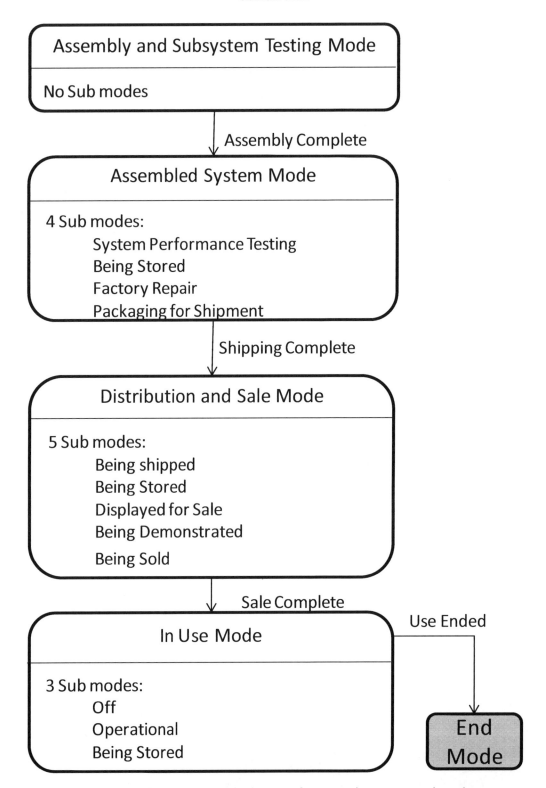

Figure 6-10 An example top level life cycle mode diagram for a simple commercial product.

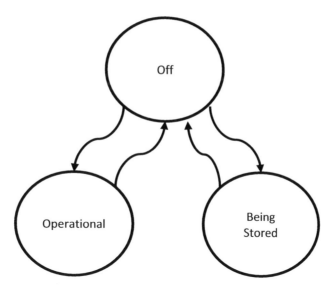

Figure 6-11 A mode transition diagram for a simple commercial product in its IN USE mode.

6.3.3.3 Kano Diagrams - A **Kano diagram** is an example of a tool that takes very little of a team's time but sometimes has a huge payoff. A Kano diagram categorizes system characteristics in the three categories of **Must Be, More is Better** and **Delighters**. Each category has a particular behavior on a plot of customer response vs. the degree to which the characteristic is fulfilled. An example Kano diagram is shown in Figure 6-12. The customer response ranges from dissatisfied to neutral to delight. Characteristics that fit the More is Better category fall on a line from dissatisfied to delight depending on the degree to which the characteristic is fulfilled. Characteristics that are classified as Must Be are those the customer expects even if the customer didn't explicitly request the characteristic. Thus these characteristics can never achieve a customer response above neutral. Characteristics that fit the Delighter category are usually characteristics the customer didn't specify or perhaps didn't even know were available. Such characteristics produce a response greater than neutral even if only slightly present. Obviously characteristics can be displayed in a three column list as well as a diagram.

The reasons to construct a Kano diagram are first to ensure that no Must Be characteristics are forgotten and second to see if any Delighters can be identified that can be provided with little or no impact on cost or performance. Kano diagrams are not intended to be discussed with customers but to assist the development team in defining the best value concept. The "Trend with Time" arrow on this diagram is not part of a Kano diagram but is there to show that over time characteristics move from Delighters, to More is Better to Must Be's. The usual example used to illustrate this trend with time is cup holders in automobiles. They were Delighters when they first appeared, then they became More is Better and now they are Must Be's. The "characteristics" included in a Kano diagram may be functions, physical design characteristics, human factors, levels of performance, interfaces or modes of operation. Thus this simple tool may contribute to many of the 15 IEEE tasks defined in Figure 6-5.

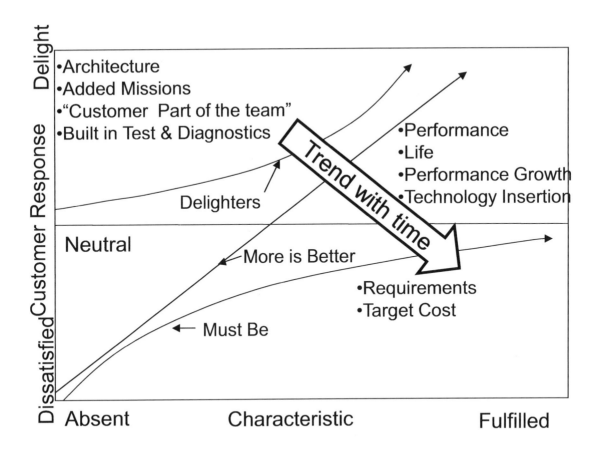

Figure 6-12 An example of a Kano diagram that classifies system characteristics into categories of Must Be, More is Better, and Delighters.

6.3.4 Define Performance and Design Constraint Requirements - Tasks 11, 13 and 14 in Figure 6-5 are related to performance and design characteristics and are discussed together here. During the analysis of market data, customer documentation and user needs, and in constructing P-Diagrams and Quality Function Deployment (QFD) QT-1 tables the top level system performance parameters and key design characteristics such as size, mass, power and data rates are identified. These parameters must be quantified and metrics must be defined that enable the team, management and customers to track progress toward meeting these important requirements. It is also necessary to judge which design characteristics are rigid constraints and which can be modified in design trade studies if necessary.

6.3.4.1 Define Tracking Metrics (KPPs & TPMs) - Tracking metrics include **Key Performance Parameters** (KPPs) and **Technical Performance Metrics** (TPMs). KPPs are critical items that the customer must have and therefore must be included in the design. TPMs provide visibility into predicting the success in meeting KPPs. TPMs may not initially meet required performance but the values should be within a range that leads to meeting the required values with further design. TPM targets may be adjusted during design as the design matures. Developing a consistent process for constructing and updating KPPs, TPMs and any other tracking metrics needed is essential to providing confidence to the system development team and to their managers and customers that work is progressing satisfactorily or if critical requirements are at risk. It is good practice, as defined by IEEE task 11, to begin the selection of tracking metrics and setting up tracking procedure during the requirements analysis task even though the design work is not yet dealing with physical design. One widely used process is presented here. Only TPMs are discussed but the process can be used for other metrics as well. The benefits of TPMs include:

- Keeping focus on the most important KPP requirements (often called **Cardinal Requirements**).
- Providing joint project team, management and customers reliable visibility on progress.
- Showing current and historical performance Vs. specified performance
- Providing early detection and prediction of problems via trends
- Being a key part of a design margin management process
- Monitoring identified risk areas
- Documenting key technical assumptions and direction

The steps in generating TPMs include:

1. Identify critical parameters to trend (e.g., from a QT-1)
2. Identify specification & target values and the design margin for each parameter
3. Determine the frequency for tracking and reporting the TPMs (e.g. monthly)
4. Determine the current parameter performance values by modeling, analysis, estimation or measurement (or combination of these)
5. Plot the TPMs as a function of time throughout the development process
6. Emphasize trends over individual estimates
7. Analyze trends and take appropriate action

Some systems have a very large number of TPMs and KPPs. Rather than report every metric in such cases develop a summary metric that reports the number or percent of metrics within control limits and having satisfactory trends. Metrics not within control limits or having unsatisfactory trends are presented and analyzed for appropriate actions. This approach works well if those reviewing the metrics trust the project team's process.

It is recommended to track and report both **margin** and **contingency** with the **current best estimate** (CBE). Unfortunately there are no standard definitions for these parameters. Example definitions are shown in Figure 6-13 where contingency is the amount of unallocated value being held by engineers to be used if necessary to meet the specified value and margin is the difference between the combined allocation with contingency and the specified value.

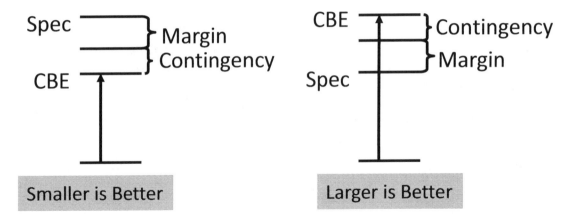

CBE = Current Best Estimate
Contingency represents 1 σ uncertainty in CBE

Figure 6-13 Example definitions for margin and contingency where contingency is defined as the current best estimate plus/minus one sigma.

An example of a TPM tracked monthly for a characteristic for which smaller-is-better, e.g. mass or power, is shown in Figure 6-14 with the plotted points being the CBE plus contingency.

Included in the chart is the specification value, shown as a line, a target value selected by the project team and plotted as a point at the planned end of the development cycle, a margin depletion line, upper and lower control limits as well as the plotted values of CBE plus a contingency. In some cases the upper and lower control limits are selected to converge as they near the target value. Selecting the target value, slope of the margin depletion line and the control limits is based on the judgment of the project team.

Smaller-is-better characteristics like mass and power consumption tend to increase during the development so initial values should be well under the specification value. Thus a reasonable strategy for selecting the margin depletion line is to link the initial estimate with the final target value. Control limits are selected to provide guidance for when design action is needed. If the characteristic exceeds the upper control limit design modifications may be needed to bring the characteristic back within the control limits. If the characteristic is below the lower control limit then the team should look for ways that the design can be improved by allowing the characteristic to increase toward the margin depletion line value at that time. For example, in Figure 6-13 the first three estimates trend to a point above the upper control limit. The fourth point suggests that the project team has taken a design action to bring the characteristic back close to the margin depletion line.

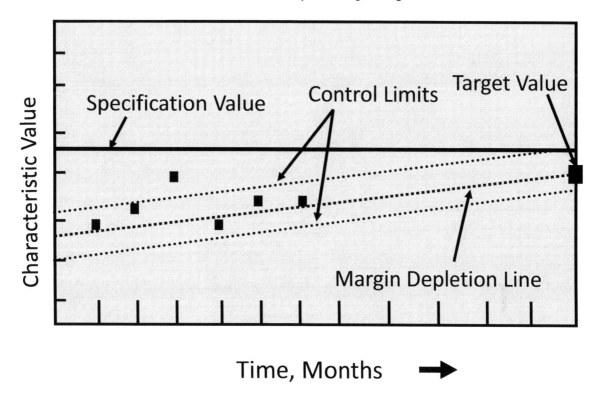

Figure 6-14 An example TPM chart for a characteristic for which smaller-is-better.

Many requirements must be broken down into separate components, attributable to each contributing factor; e.g. mass of a system can be broken down into the mass for each subsystem. TPMs for such requirements are a rollup of the contributing components. A variation on a TPM particularly useful for such requirements is to explicitly include the **basis of estimate** (BOE) methods of arriving at the CBE value each period. Typical BOE Methods are **allocation, estimation, modeling** and **measurement**. **Allocation** is the process of taking a top-level roll-up and budgeting a specified amount to each contributing component. **Estimation, modeling** and **measurement** are methods applied to each contributing component and then combining them to determine the result for the top level requirement. Figure 6-14 is an example of a smaller-is-better characteristic with the CBE shown as a stacked bar with the proportion of the CBE from each method explicitly shown.

The proportions allocated and estimated decrease as modeling and measurements are made, which increases the confidence in the reported CBE value. In the example shown the CBE is the sum of values for each contributing component with different uncertainties and the plotted values, i.e. the top of the stacked bar are the CBE plus the RMS of the uncertainties, i.e. the contingency in this case. For a larger-is-better characteristic the plotted value is the CBE minus the RMS of the uncertainties assuming the definitions in Figure 6-13.

One alternative method of tracking and reporting metrics is using spreadsheet matrices rather than charts. This approach allows the values for each contributing component to be tracked and reported. An example for a smaller-is-better characteristic attributable to three component subsystems is shown in Figure 6-16.

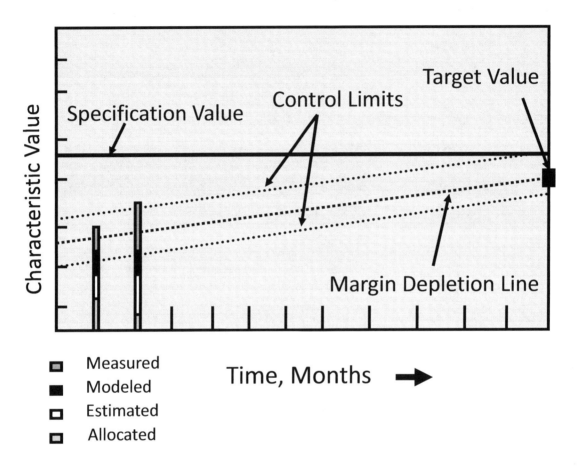

Figure 6-15 An example of a TPM chart where the plotted values include the basis of estimate methods to provide more information about the confidence in reported values.

The total allocated value can be either the specification value or the specification value minus a top level margin, i.e. a target value. The BOE methods for each contributing component CBE can be included in additional columns if desired. The disadvantage of a matrix format compared to a time chart is that it isn't as easy to show trends with time, margin depletion line value and control limits. Matrices are useful for requirements that involve dozens of contributing components; each important enough to be tracked. The detail can be presented in a matrix with color coding to highlight problem areas and either a summary line or summary chart used for presentations to management or customers.

A third alternative for tracking metrics is a tree chart. An example using the data from Figure 6-16 as mass in Kg is shown in Figure 6-17.

A tree chart has similar disadvantages as a matrix compared to a time chart but both trees and matrices have the advantage of showing the values for contributing components. Sometimes it is valuable to use both a time chart and a tree or matrix for very important TPSs or KPPs.

	CBE	Contingency	Allocation	Margin
Subsys 1	1.4	0.2	1.7	0.1
Subsys 2	2.8	0.3	3.5	0.4
Subsys 3	0.9	0.1	2	1
Total	5.1	0.4	7.2	1.7

Figure 6-16 An example TPM in matrix format for a smaller-is-better characteristic attributable to subsystems.

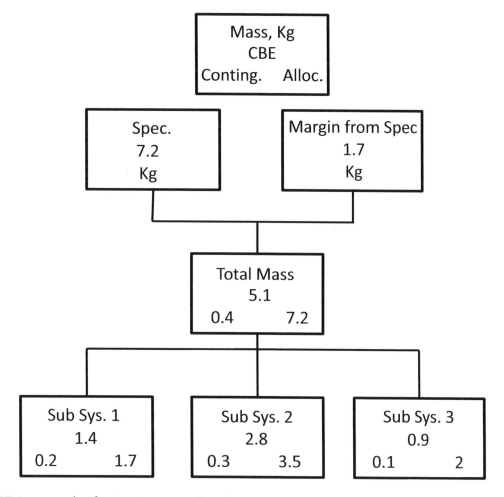

Figure 6-17 An example of a TPM in a tree chart format.

Common sense guidelines to follow in managing tracking metrics include:

- Keep metrics accurate and up to date
- Be flexible with metric categories (add, delete, modify as development progresses)
- Dovetail into customer process
- Plan ahead to achieve maturing Basis Of Estimates (BOE)
- Make performance prediction part of normal design process (flow-up, not run-down)
- Include in program status reports, e.g. monthly.

It should be part of each IPT's normal work to report or post to a common database the estimates for metrics under their design responsibility. This makes it easy for one of the system engineers to compile up to date and accurate metrics for each reporting period. If this is not standard practice then someone has to run down each person responsible for part of a metric each period to obtain their input; a time consuming and distasteful practice.

6.3.4.2 Spider Diagrams - Another useful diagram is the **spider diagram**; named from its shape. Spider charts are useful for tracking progress of requirements that are interrelated so that careful attention is needed to avoid improving one at the expense of others. An "**ility**" diagram is a good example of a spider chart that tracks progress toward goals or requirements for selected ilities; i.e. reliability, maintainability, reparability, etc. An ility diagram treats the system or product as a single entity. It is constructed by selecting and ranking in importance eight ilities as described by Bart Huthwaite in his book *"Strategic Design"* cited earlier[2-3]. Initially measures and goals are established for each ility. Later, as the system design takes form estimates or calculations of each ility are made and plotted on the diagram. As system design trades progress and the design is refined the progress toward goals in tracked. By including the most important eight ilities and tracking their measures on a chart the design team has a clear picture of how well the design is progressing and if any ility is being sacrificed for the benefit of some other ility or performance parameter. An example ility chart as it might appear in the middle of system trades is shown in Figure 6-18. Here the eight ilities are numbered in rank order of importance so that if tradeoffs cannot be avoided the most important ilities are favored over less important one. The line connecting the eight legs is the CBE for each ility. Trends can be indicated by keeping the previous three or four CBEs on the diagram.

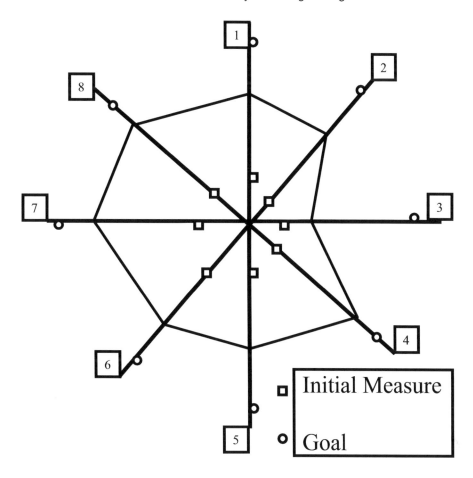

Figure 6-18 An example of the form of an "ility" chart that tracks progress toward goals for eight ilities ranked in importance from 1 to 8.

6.3.5 Balancing Customer Needs - A major goal of requirements analysis is to ensure that the set of requirements are complete, accurate, and non-redundant. An additional activity during the requirement analysis is to identify opportunities to optimize the customer's desires and needs. This task is considered part of the **Technical Management** task shown in Figure 6-4. Working with the customer to adjust requirements often presents an opportunity to save cost or schedule with little impact on the customer. This may also provide significant improvements in a highly desired performance parameter with little impact on a lesser desired parameter. These changes are best identified during the requirements analysis. As the program moves into the architecture, then design phase, changes become more costly providing less opportunity for savings. The understanding that a very large percentage of the program cost and schedule are established very early in the program is an important responsibility of the system engineer. Requirements drive the system architecture trades, and once the architecture is defined, the opportunity for savings is a small percentage of the opportunity early in the program. Pugh diagrams (described in Chapter 8) and QFD tables (described in Chapter 7) are valuable tools for assessing these opportunities.

6.4 Functional Analysis and Allocation

Figure 6-5, the list of 15 tasks, shows that **Functional Analysis** and **Allocation** are necessary to accomplish subtask 12, Define Functional Requirements. **Functional analysis** decomposes each of the high level functions of a system identified in requirements analysis into sets of lower level functions. The performance requirements and any constraints associated with the high level functions are then **allocated** to these lower level functions. Thus the top level requirements are flowed down to lower levels requirements via functions. This decomposition and allocation process is repeated for each level of the system. The objective is to define the functional, performance and interface design requirements. The result is called the **Functional architecture** and the collection of documents and diagrams developed is the **Functional view**.

A **function** is an action necessary to perform a task to fulfill a requirement. Functions have inputs and outputs and the actions are to transform the inputs to the outputs. Functions do not occupy volume, have mass, nor are they physically visible. Functions are defined by action verbs followed by nouns; for example, condition power or convert frequency. A complex system has a hierarchy of functions such that higher level functions can be decomposed into sets of lower level functions just as a physical system can be decomposed into lower level physical elements. A higher level function is accomplished by performing a sequence of sub-functions. Criteria for defining functions include having simple interfaces, having a single function each, i.e. one verb and one noun; having independent operations and transparency, i.e. each does not need to know the internal conditions of the others. There are both explicit and implicit or derived functions. Explicit functions are those that are specified or decomposed from specified functions or performance requirements. Implicit or derived functions are those necessary to meet other specified capabilities or constraints.

Sometimes inexperienced engineers ask why they have to decompose and allocate functions for design elements at the subsystem or lower levels; they believe once they know the top level function, the performance requirements and any constraints defined in the requirements analysis task they can design the hardware and software without formal functional analysis/allocation. One simple answer is that items like time budgets and software lines of code are not definable from hardware or software; they are defined from the decomposed functions and sequences of functions. This is because hardware or software does not have time dimensions, only the functions the hardware or software performs have time attributes. Similarly the question of how many lines of code or bits of memory only have meaning in terms of the functions the software code is executing. Other reasons become more apparent as we describe functional design and design synthesis.

A diagram, simplified from Figure 6-4 and shown in Figure 6-19, helps understand both the question asked by inexperienced engineers and the answer to their question.

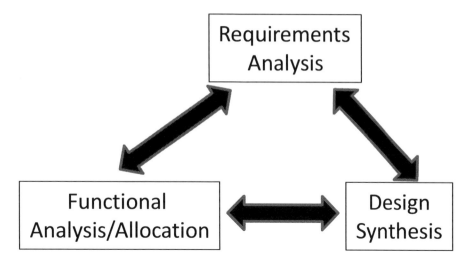

Figure 6-19 Developing a hardware/software system design is iterative and has multiple paths.

Figure 6-19 illustrates that although the primary design path is ultimately from requirements to hardware/software design; primary because when the design is complete the resulting hardware/software design fulfills the requirements, other paths are necessary and iteration on these paths is necessary to achieve a balanced design. The paths between requirements and functional design and between functional design and hardware/software design are necessary to:

- Validate functional behavior
- Plan modeling and simulation
- Optimize physical partitioning
- Facilitate specification writing
- Facilitate failure analysis
- Facilitate cost analysis and design to cost efforts
- Facilitate concept selection
- Define software budgets and CON-OPS timelines.

Although the path from requirements analysis to design synthesis isn't formally shown in Figure 6-4 it is used when a team elects to design a product around a particular part, such as a state of the art digital processor or new and novel signal processing integrated circuit chip. However, having preselected a part doesn't negate the need to define the functions performed by the part and verify that the performance requirements and interfaces are satisfied.

6.4.1 Decompose to Lower-level Functions – Decomposition is accomplished by first arranging the top level functions in a logical sequence and then decomposing each top level function into the logical sequence of lower level functions necessary to accomplish the top level functions. Sometimes there is more than one "logical" sequence. It is important to examine the decomposed functions and group or partition them in groups that are related logically. This makes the task of allocating functions to physical elements easier and leads to a better design, as will be explained in a later section. When more than one grouping is logical then trade studies are needed. Although the intent is not to allocate functions to physical entities at this point functions should not be grouped together if they obviously belong to very different physical elements. The initial grouping should be revisited during the later task of allocating functions to physical elements and this process is described in more detail in a following section.

The DoD SEF has an excellent description of functional analysis/allocation and defines tools used to include **Functional Flow Block Diagrams** (FFBD), **Time Line Analysis,** and the **Requirements Allocation Sheet**. N-Squared diagrams may be used to define interfaces and their relationships. Spreadsheets are also useful tools and are used later in various matrices developed for validation and verification. A spreadsheet is the preferred tool for developing a functions dictionary containing the function number, the function name (verb plus noun) and the detailed definition of each function and its sub functions. The list of function numbers and names in the first two columns of the functions dictionary can be copied into new sheets for the matrices to be developed in the design synthesis task.

Spreadsheets do not lend themselves to identifying the internal and external interfaces as well as FFBDs so the FFBD is the preferred tool for decomposition. Time Line Analysis and the Requirements Allocation Sheet are well described in Supplement 5 of the DoD SEF and need no further discussion here. Similarly the **Integration Definition for Function Modeling** (IDEF0) is a process-oriented model for showing data flow, system control, and the functional flow of life cycle processes. It is also well described in Supplement 5 of the DoD SEF. The collection of documents, diagrams and models developed using all of the tools is the **Functional view**. The functional architecture is the FFBDs and time line analyzes that describe the system in terms of functions and performance parameters.

Although the FFBD is discussed in the DoD SEF there are some additional important aspects of this tool that are covered here.

6.4.1.1 Functional Flow Block Diagrams - Graphical models are used to define and depict the sequence of functions making up a higher level function necessary to fulfill requirements. Functional Flow Block Diagrams (FFBD) can be **process-oriented models** that represent functions as nodes and objects as the connecting links. Nodes are typically labeled but links are typically unlabeled in process-oriented FFBDs. Figure 6-20 is a FFBD for two of the functions of a digital camera. Here only one external interface is identified, i.e. the interface with light from the scene.

Part of the task of functional design is to decompose high level functions into the lower level functions necessary to carry out the action implied by the high level function. For example, the function "image scene" can be decomposed as shown in the FFBD of Figure 6-21. In this example the objects linking the four lower level functions are also labeled. Note that nodes are numbered as well as named with the numbers indicating the level of the functions in the overall hierarchy of functions. The numbers are selected to provide traceability; e.g. sub functions decomposed from a function 1.1 with n sub functions are numbered 1.1.1 to 1.1.n.

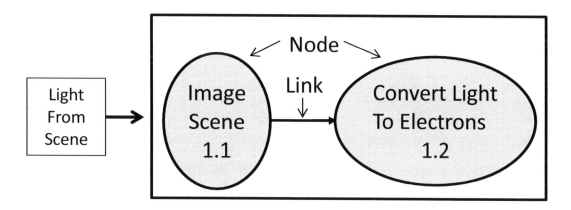

Figure 6-20 A FFBD developed as a process-oriented model has named and numbered functions as nodes and objects as links.

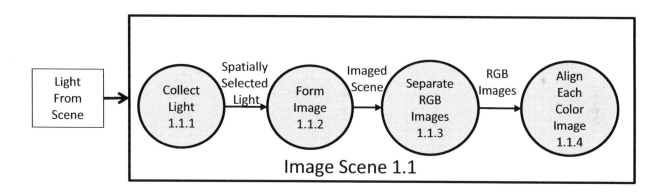

Figure 6-21 The function Image Scene 1.1 from Figure 6-20 can be decomposed into four lower level functions.

Notice that functions 1.1.2 and 1.1.3 in Figure 6-21 could be interchanged in the sequence and still be a logical sequence. This is a simple example of having more than one logical sequence of lower level functions. The "best" sequence is determined by conducting design trade studies of the choices or when these functions are allocated to physical elements. Chapter 8 describes methods for selecting the "best" sequence. Note also that the functions 1.1.1 and 1.1.2 are easily grouped whereas 1.1.1 and 1.1.3 or 1.1.2 and 1.1.3 are not easily grouped. Therefore the sequence shown is likely to be the preferred sequence, at least until design trade studies are complete.

An alternative to the process-oriented model of a FFBD is an **object-oriented model** with the nodes and links reversed. An example of an object oriented model is shown in Figure 6-22.

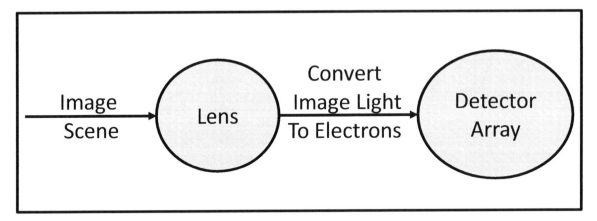

Figure 6-22 An example of a FFBD as an object-oriented model of two of the functions of a digital camera with named functions as links and named objects as nodes.

The decision to develop process-oriented or object-oriented models depends upon the experience of the systems engineer and the details of the system being designed. A comparison of the two approaches resulting from an analysis of both approaches by Doyle and Pennotti[6-2] is shown in table 6-1.

As with context or domain diagrams it is helpful to have pattern diagrams for all levels of FFBDs representing the class of systems that includes the system being designed. This saves considerable time compared to having to generate the FFBDs for all the levels of a new system under development. The objective is that the pattern diagrams contain all possible functions, sub functions and external and internal interfaces of the entire class of systems. Then the task becomes examining each top level function and interface to determine if it belongs to the new system. If a function does not belong it is deleted along with all sub functions decomposed from the unneeded top level function. After deleting all unnecessary functions and interfaces from the top level and the sub functions and interfaces traceable to the deleted top level functions and interfaces it is necessary to examine the sub functions in each level to ensure that only necessary ones are kept. The system being designed may include a top level function from the pattern but may not include the entire set of sub functions decomposed from the top level function. It is also necessary to examine the partitioning and grouping of functions as the best choice is likely to be system design specific and not necessarily that captured in the pattern diagrams.

Object-Oriented	Process-Oriented
Easier to understand how the system does what is required without having to understand the components.	Easier to understand the roles and responsibilities of the components without having to understand the system.
Provides a way to manage a project when implementation trade-offs are unresolved when the project is divided among work groups.	Provides a way to manage a project by reducing dependencies between software components that are being developed by different work groups.
Provides an implementation independent view of the system when many implementations are possible and it is important to keep all options open.	Provides well understood, standardized implementations that maximize reuse of design patterns, code and skills when functions have been allocated to software.
Easy to trace requirements because inputs and outputs are consistent between levels of decomposition.	Changes have less ripple effect because the inputs and outputs are hidden between levels of decomposition.

Table 6-1 Selecting an Object-Oriented or Process-Oriented model depends on the details of the specific system design.

6.4.2 Allocate Performance and Other Limiting Requirements - It is important not to get caught up in the process of developing the various documents and diagrams and lose sight of the objective that is to develop a new system and that a primary responsibility of the systems engineers is to define complete and accurate requirements for the physical elements of the new system. Having decomposed the top level system modes into their constituent modes and the top level functions of the system into the lower level functions required for each of the decomposed modes the next step is to allocate (decompose) the performance and other constraining requirements that have been allocated to the top level functions to the lower level functions.

The primary path is to follow the FFBDs so that requirements are allocated for every function and are traceable back to the top level functional requirements. Traceability is supported by using the same numbering system used for the functions. Requirements Allocation Sheets may be used, as described in the DoD SEF or the allocation can be done directly in whatever tool is used for the Requirements Database. Other useful tools are the scenarios, Timeline Analysis Sheets (TLS) and IDEFO diagrams developed during requirements analysis and functional decomposition.

If the team followed recommended practice and began developing or updating applicable models and simulations these tools are used to improve the quality of allocated requirements. For example, budgeting the times for each function in a TLS based on the results of simulations or models is certainly more accurate than having to estimate times or arbitrarily allocate times for each function so that the time requirement for a top level function is met.

Another example is any kind of sensor with a top level performance requirement expressed as a probability of detecting an event or sensing the presence or absence of something. This type of performance requirement implies that the sensor exhibit a signal to noise ratio in the test and operational environments specified. Achieving the required signal to noise ratio requires that every function in the FFBD from the function that describes acquiring the signal to the final function that reports or displays the processed signal meets or exceeds a level of performance. Analysis either by models or simulations is necessary to balance the required performance levels so that the top level performance is achieved with the required or desired margin without any lower level functions having to achieve performances that are at or beyond state of the art while other functions are allocated easily achievable performances.

Functional trees are very useful for budgets and allocations, particularly Concept of Operations (Con-Ops) timelines and software budgets since physical elements don't have time, lines of code (LOC) or memory requirements but functions do. Transforming the FFBD into an indented list of numbered and named functions on a spreadsheet facilitates constructing a number of useful tables and diagrams. Consider a timeline analysis sheet (TLS) for a hypothetical system having two functions decomposed as shown in Figure 6-23.

Function			Time			
Number	Name	Seconds	2	4	6	8
1	Function 1	5				
1.1	Function 1.1	0.5				
1.2	Function 1.2	1.8				
1.2.1	Function 1.2.1	0.75				
1.2.2	Function 1.2.2	1.5				
1.3	Function 1.3	1.3				
2	Function 2	3				
2.1	Function 2.1	1.5				
2.2	Function 2.2	1.5				
2.3	Function 2.3	1.5				
2.4	Function 2.4	1				

Figure 6-23 A hypothetical TLS for a system with two functions decomposed into its sub functions.

The TLS illustrates both the time it takes to execute each sub function in a particular Con-Ops scenario and the starting and stopping times for each time segment. If the functions were to be executed sequentially nose to tail then just the numerical time column would be needed and the total time would be determined by the sum of the individual times.

The same function list can be used for software budgets or allocations. An example is shown in Figure 6-24.

Function		Software Budget	
Number	Name	LOC	Memory (Kbytes)
1	Function 1		
1.1	Function 1.1	82	1
1.2	Function 1.2		
1.2.1	Function 1.2.1	23	0.6
1.2.2	Function 1.2.2	65	0.4
1.3	Function 1.3	28	2
2	Function 2		
2.1	Function 2.1	13	12
2.2	Function 2.2	41	10
2.3	Function 2.3	32	30
2.4	Function 2.4	16	10
	Sum	300	66
	Margin	300	66
	Percent Margin	100%	100%
	Totals	600	132

Figure 6-24 Software lines of code and memory can be budgeted or allocated to a list form of the system functions.

6.4.3 Define and Verify Functional Interfaces (Internal/External) - Logical interfaces with external elements are defined in the context diagram and the FFBDs and internal interfaces are defined in the FFBDs. Both types of interfaces must be analyzed to verify that all interfaces are properly located and defined. Examine each external interface and verify that the information coming from or going to the interface matches the information being handled by the parent function in the chain of lower level child functions. Similarly examine each function and verify that all information coming from or going to the function is accounted for; that no function has an output that doesn't go to either another function or to an external logical interface; and that no function requires information that is not coming to the function from another function or external interface. This task is made easier if the links in a process-oriented FFBD are labeled. An example of a simple process-oriented FFBD of a toaster with internal and external interfaces is shown in Figure 6-25.

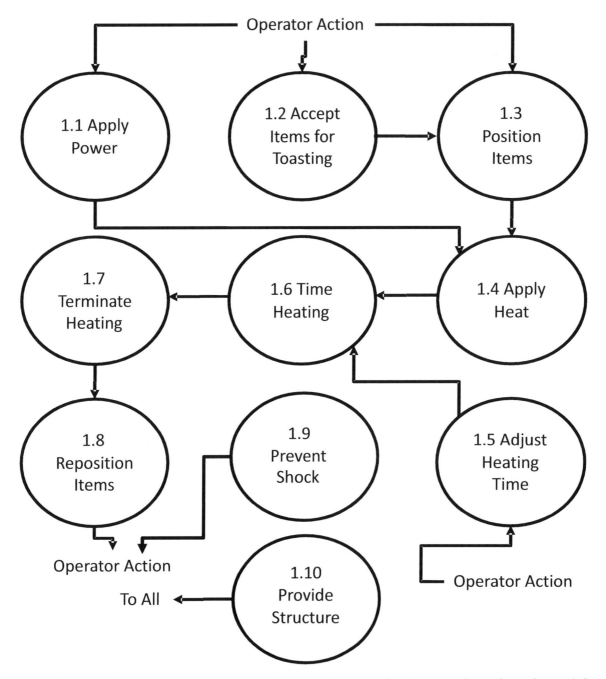

Figure 6-25 An example of a FFBD for a toaster showing the internal and external interfaces for each function of the Operational mode.

(Apologies to experienced designers of toasters for any mistakes by the authors who have limited domain knowledge of toaster design. We use the example of a toaster because it is simple enough that diagrams and models fit on a page and everyone has some idea of what a toaster does and how it might work. To those "virtuous and pure" engineers whose response is "toasters don't apply to my work so these examples are useless to me" we remind you that the authors have used these same methods on systems costing hundreds of millions of dollars to develop. Learn the methodologies illustrated by these examples and don't be put off by errors or incompleteness in these examples or the fact that your systems are much more complex.)

A "from" "to" matrix of functions in a particular mode is an alternate tool for defining interfaces for functions. An example is shown in Figure 6-26 for a toaster in its operational mode.

To Functions for Operational Mode	From Functions for Operational Mode										
	1.1 Apply Power	1.2 Accept Items for Toasting	1.3 Position Items	1.4 Apply Heat	1.5 Adjust Heating Time	1.6 Time Heating	1.7 Terminate Heating	1.8 Reposition Items	1.9 Prevent Shock	1.10 Provide Structure	External Interface / Operator Action
1.1 Apply Power											x
1.2 Accept Items for Toasting											x
1.3 Position Items		x									x
1.4 Apply Heat	x		x								
1.5 Adjust Heating Time											x
1.6 Time Heating				x	x						
1.7 Terminate Heating						x					
1.8 Reposition Items							x				
1.9 Prevent Shock											
1.10 Provide Structure	x	x	x	x	x	x	x	x	x		
External Interface / Operator Action								x	x		

Figure 6-26 A Matrix of Functions to Functions is an alternate tool for defining internal and external interfaces among functions.

N-Squared diagrams are useful tools for analyzing interfaces for systems with functions having many internal interfaces. This tool also provides verification of the grouping and sequencing of lower level functions. It's much easier to detect sequencing problems in an N-Squared diagram than on a FFBD. An example of an N-Squared diagram used for defining internal and external interfaces is shown in Figure 6-27. The advantages of the N-Squared diagram aren't apparent in this simple case but imagine if the functions were more randomly sequenced along the diagonal. Then there would be arrows on the left of the diagonal indicating poor sequencing.

It is good practice to develop two different tools for defining internal and external interfaces; for example a FFBD and an N-Squared diagram. The two are then compared to verify that all interfaces are defined, grouped and sequenced correctly and consistent with the definitions of functions in the data dictionary. The small amount of time it takes to verify functional interfaces via two different tools is sound risk mitigation against making a mistake that isn't discovered until system or subsystem testing when correcting the error is very costly.

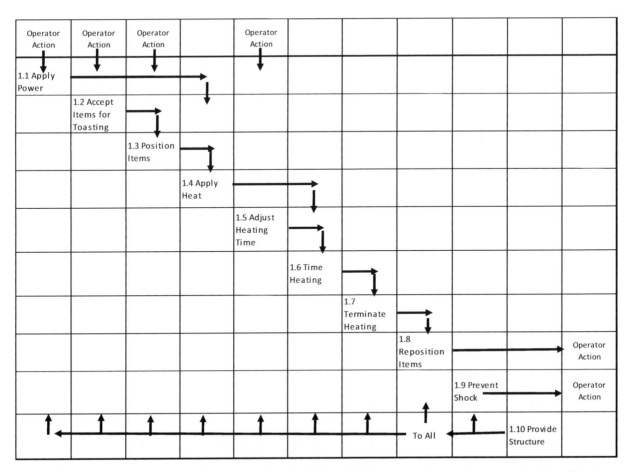

Figure 6-27 An N-Squared diagram is an excellent tool for defining, grouping and sequencing interfaces.

6.4.4 Verify the Functional Architecture - The **functional architecture** is the FFBDs and the allocated requirements. The collection of all documentation developed during the functional analysis and allocation task is called the functional view. The final task in defining the functional architecture is to review all of the functional view documentation for consistency and accuracy. Check the functions defined for each mode and sub mode to verify that no functions are missing and that the requirements allocated to each function are appropriate for each mode and sub mode. An example of a matrix of modes to functions useful for verifying that all top level functions needed for each sub mode of a toaster in its In Use Mode are defined properly is shown in Figure 6-28. This example only examines one system mode but the process for examining all modes and all lower level functions is just an extension of the matrix.

The methodology of verifying functional design by using two different tools to describe the same functional requirements also applies to mode transitions. Figure 6-29 is an example of a matrix used to define the allowed transitions among modes of the In Use mode as previously defined with a mode transition diagram in Figure 6-11. Although this is a trivial example it illustrates the methodology.

System Functions		System Modes for In Use Mode		
		Off	Operational	Being Stored
1.1	Apply Power	x	x	
1.2	Accept Items for Toasting		x	
1.3	Position Items		x	
1.4	Apply Heat		x	
1.5	Adjust Heating Time	x	x	
1.6	Time Heating		x	
1.7	Terminate Heating		x	
1.8	Reposition Items		x	
1.9	Enable Cleaning	x		
1.10	Store Power Cord	x		x
1.11	Prevent Shock	x	x	x

Figures 6-28 An example of a Functions to System Modes matrix that facilitates verifying that all functions are defined for all modes.

		Modes of In Use Mode		
		Off	Operational	Being Stored
Modes	Off		x	x
	Operational	x		
	Being Stored	x		

Figure 6-29 Allowable mode transitions can be defined in a matrix as well as in a diagram.

Revisit the documentation in the operational view to verify that the functional architecture accounts for every function necessary to fulfill the operational requirements and that no unnecessary functions have been added. Verify that every top level performance and constraining requirement is flowed down, allocated and traceable to lower level requirements and that there are no lower level requirements that are not traceable to a top level requirement.

6.5 Developing the Requirements Database

The objective of Requirements Analysis and Functional Analysis/Allocation is to facilitate establishing the **requirements database** as indicated in Figure 6-5 Developing requirements and writing specification documents are skills that are often taken for granted. Good requirements and clearly written specifications are critical to the overall success of a program, thus need to be developed by a team with an understanding of the implications of requirements downstream in the program. Every program has its own unique aspects of requirements developed from the understanding of the functionality, performance, and interface needs. However, there are a number of characteristics that are applicable to all requirements.

Requirements must be achievable. As requirements are finalized before the architecture is complete, the systems engineer must have enough understanding of the concept to have confidence that when the design is complete, the specified performance is achieved. A preliminary model of the system often provides confidence in the projected performance. However, the model likely includes assumptions about elements that have not been designed. This is where the experience of the systems engineer is invaluable in being able to ensure the assumptions are reasonable.

Requirements must be complete. There should be no assumptions made about the final product. A good cross check is once all requirements are complete, ask the question, 'if the system is delivered accomplishing the stated requirements, and does nothing else, will the system do what is desired'. A common pitfall is when the team generating the requirements knows too much about what is desired. When the team has too much knowledge, the requirements often include more detail than desired and move from being logical, to specifying implementation. This often results when developing an upgrade to a legacy product.

Requirements should not be redundant. Every requirement has a cost associated with it so redundant requirements drive program cost and should be avoided. Redundant requirements often result from stating characteristics that are not independent or from two different perspectives. For example, there may be requirements for total mass and total size. A separate requirement for average density is redundant and adds no value. Additionally, redundant requirements often constrain the design more than necessary, eliminating possible implementations that may reduce cost or schedule.

Requirements must be verifiable. If a requirement is written such that it cannot be verified, there is no way of determining if the delivered system is compliant. Defining the verification criteria when the requirement is written provides confidence that the requirement is verifiable. A good characteristic for avoiding requirements that cannot be verified is to make sure requirements are bounded. Requirements that include terms such as maximize, minimize, or optimize should be avoided.

Requirements must be clear. Requirements that can be interpreted to have multiple meanings, particularly by customers that may not have an engineering background, often result in the customer not getting what they expect. Minimizing the requirement to make it as concise as possible usually makes it more understandable. Here again, defining verification compliance criteria with the requirement helps ensure the requirement is understood. Including diagrams, tables and other labeled graphical models where applicable helps eliminate ambiguity.

Requirements should be identified with the use of the word 'shall'. Use of the word shall is to be reserved for requirements. Every 'shall' statement is to be identified as a deliverable requirement that must be verified. If an item is not to be verified, it should not include a 'shall' statement. Any statements including terms such as should, desire, or goal, are not requirements and are not verified. These statements may convey customer intent but are not requirements. If alternatives are being considered that may address the customer desires, they may be considered, but they should not drive more costly alternatives.

Requirements include a single performance item to be verified. Complex requirements stating multiple elements are to be eliminated. Requirements including the word 'and' often include more than one performance statement and should be avoided.

Requirements should avoid negative statements. Negative requirements often define a broad range of conditions to be avoided and are often impossible to verify. A typical negative requirement might state 'the system shall not interfere with any other systems'. This requirement has potentially an infinite number of items to be verified.

Requirements have a typical structure of 'The ***object*** shall provide ***performance*** ... modifying conditions'. This structure clearly defines what the requirement applies to, what performance is to be delivered, and if there are any conditions that apply.

6.5.1 Organizing the Requirements Database - Writing proper specification documents is labor intensive work even when using modern tools designed for requirements management. When developing specifications, organizing requirements within the specifications, as well as their relationship between specifications, is important.

It is very beneficial to the development team for all specifications to follow a common structure. There are many recommendations for organizing specification with the most popular being Mil standards, IEEE standards, and INCOSE standards. These standards are equally valid and should be selected based on the best fit with the customer. Having selected an applicable standard, additional detail should be developed to elaborate on what information is included in each section. This defined structure should then be followed for all specifications on the program.

Regardless of what structure is selected, a common theme to all organizations is to organize performance requirements by system functions. As previously described, system analysis starts with the development of the context diagram, followed by the identification of the functions. All performance should be associated with the identified function. If a performance is identified that is not associated with a function, it is likely that one is missing from the defined set of functions. The association of requirements to functions is important when trading off performance during the design phase. This association allows the customer to fully understand the true impact to system operation when individual requirements performance is adjusted.

Having a common defined structure for specifications benefits both the individual specifications, as well as the relationships of requirements between specifications. The commonly defined structure allows each engineering discipline to look for the same kinds of requirements in the same location, across all individual specifications. When decomposing requirements and establishing relationships between specs, the common structure is also a benefit as requirements tend to be decomposed in lower level specs into the same sections when the parent requirement is located in the upper level spec.

Organizations often focus on market segments that have similar products or systems. In such cases the use of patterns and templates to provide a common structure can save both time and costs. Common market segments and similar products have a large percentage of requirements in common. Whereas the exact performance for a specific parameter may change between products, the requirements may be generally common. For example, sports cars and sedans have different performance, but both have drive train requirements such as top speed, gas mileage, etc. This concept for a broad set of common requirements can be extended to all requirements that are applicable to all vehicles. An auto manufacturer would benefit from a vehicle spec pattern that includes these requirements.

The idea is to organize the top level specification and decomposed specifications so that the parent to child relationships are linked such that the links are representative of a family of products or systems. The best organization of specifications is dependent on characteristics of the product or system family but experienced systems engineers can use their judgment to choose a suitable organization. An example of a specification organization suitable to a system family with similar electronics subsystems is shown in Figure 6-30.

The customer documents are different for each different system and must be analyzed, flowed down and linked for traceability into a set of top level organization documents as illustrated in the figure. Sometimes customers require delivery of hardware or software that is not part of the system but supports the system. Examples might be special test equipment or algorithms for diagnostics or data analysis. It is unlikely that templates can be developed for such specifications included in the Figure 6-30 as "Other Specifications". Templates can be developed for the system and typical subsystem specifications, the ICD and mission/quality assurance specifications.

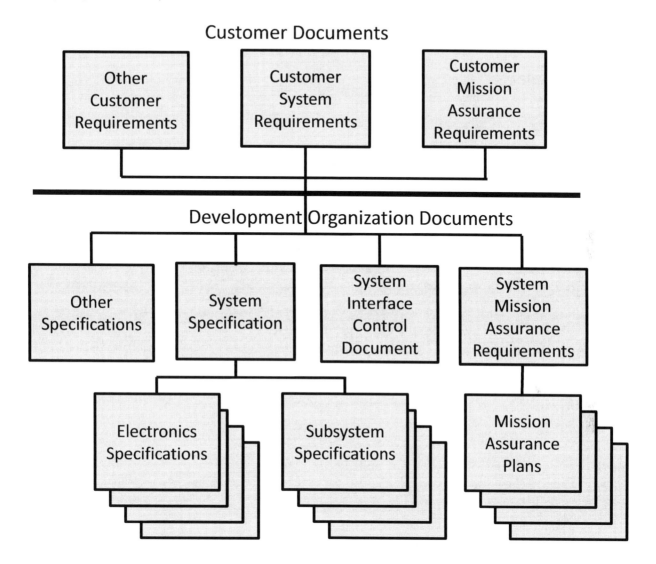

Figure 6-30 *An example specification organization that can be a template for a family of systems having similar electronic subsystems.*

Complete pattern templates include the entire specification content, including all requirement statements and links for the likely requirements of the family of systems. The requirement statements in the pattern have TBDs in place of the numeric parameters of actual values for an actual product. These TBDs are replaced with real numbers when the pattern is tailored for a real program. Converting the pattern templates to specific system specifications involves examining each statement to decide if it is a requirement for the system under development, determining the performance parameters and conditions that apply to the requirement and confirming the traceability links. If an individual statement does not apply it is deleted from the pattern template. In general, it is beneficial that if an entire section is not applicable, content can be deleted, but the title is maintained as 'Reserved', so that the numbering of statements is not changed thus maintaining a common structure across all specifications for all levels of the programs specifications. If a new requirement is needed it is added to both the new specification and to the pattern template. Thus writing the specifications becomes an editing task and takes much less time than writing the specification documents from scratch. If a new section is required, it is added at the end of the section so the existing section numbering is not modified.

The advantage of developing and using pattern templates is to help guide the development taking advantage of lessons learned on previous programs. Patterns help ensure all functions and requirements are addressed without having to rely on individuals to make sure they haven't forgot anything. Patterns should be developed with significant support of subject matter experts thus allowing programs to take advantage of their expertise even if they may not be available to support requirement development on a specific program. Although not shown in Figure 6-30 it should be obvious that the pattern template approach can be carried to other system documentation that is part of the information architecture, e.g. test plans at the subsystem and system level, maintenance plans, logistics plans etc. The more documentation that can be configured for reuse the faster the systems engineering work can be accomplished for new systems. As patterns include links to not only decomposed specifications, patterns should also be linked to and drive common integration and verification processes thereby driving integration and test reuse resulting in significant savings of time and engineering effort.

Mature organizations have well developed mission assurance plans. The detailed plans can be linked to a top level requirements document ensuring traceability of any requirement from the top level document to the appropriate detailed plan. The system engineering task is to analyze the customer's mission assurance requirements and edit the top level document to be in compliance with the customer's requirements. All customer requirements matching system requirements at the top level are already linked to appropriate detailed plans. Any new customer requirements must be added and possibly detailed plans require editing to comply with new requirements. Any requirements that are in the organization's template but not in the customer's requirements must be deleted. Having the links in place makes it easy to determine the detailed plan requirements that do not apply.

6.6 Design Synthesis

Completing the initial definition of the functional architecture sets the stage for beginning **design synthesis**. The design synthesis task defines physical elements of hardware and software to carry out the functions in the functional architecture and to fulfill the requirements allocated to the functions. It is an **allocation** and **partitioning** task. Allocation refers to assigning physical elements to the functions and partitioning refers to the grouping of functions and physical elements. It's helpful if the functional architecture is defined to the second level, at least in draft form, before beginning design synthesis. Design synthesis is done in steps. Usually the steps are called **concept design, preliminary design** and **detailed design**. Each step adds more detail to the design and defines the design to lower levels of the system hierarchy. At the completion of detailed design a complete set of procurement documentation, manufacturing drawings, detailed software descriptions and integration and test (I&T) documentation is finalized and ready for procurement of parts, manufacturing, software coding and I&T.

6.6.1 Concept Design - Concept design is emphasized here as systems engineers have a greater role in the concept design than in preliminary and detailed design. The objective of concept design is to convert the functional architecture to a physical architecture. In this process the functional architecture and the allocated requirements may be refined and other supporting documentation developed. Three outputs from design synthesis during concept design are a **physical architecture**, a **baseline design** and a **physical view** of the system.

The physical architecture is defined by a **physical block diagram** or **signal flow block diagram** that schematically illustrates the relationships and interfaces between the physical subsystems (hardware and software) that map to the functional architecture. The physical architecture is part of the **system architecture**, which includes the enabling products and services needed by the system in all of its life cycle modes. An example of a simple physical block diagram of a candidate concept design for the toaster defined by the functions shown in Figure 6-25 is shown in Figure 6-31.

Systems that involve the collection, processing and communication of signals or similar information are often better described by a signal flow diagram. A signal flow diagram is a physical block diagram that follows the system signals from initial collection to their output from the system. Modularity is often easier to visualize in signal flow diagrams than block diagrams. Signal flow diagrams are typically more complex than simple block diagrams so it's usually best to define alternative concepts and conduct trade studies using simple block diagrams. Once the final baseline concept is selected then constructing a signal flow diagram helps explain the selected design concept better than a simple block diagram.

The **baseline design** includes the functional architecture, the physical architecture, the system specification, and the ICD. The baseline design evolves with the design maturity and is the basic item under configuration management. The baseline design is a means for facilitating decision management during the three stages of design synthesis. At each stage; concept design, preliminary design and detailed design; it is good practice to force the work to a baseline design quickly and then conduct trade studies to refine the selected baseline. Otherwise too many design decisions are open at any time and control of the design work becomes difficult.

The **physical view** includes all of the diagrams, documents, models, etc. that describe how the system is constructed, how it interfaces with humans and other supporting systems during the life cycle modes, any customer supplied equipment, and any constraints on the design or operations.

6.6.2 Decision Management during Design - The degrees of freedom of a design are greatest and the cost to make a design change is lowest during concept design. This is illustrated schematically in Figure 6-32. The high degrees of design freedom means that design alternatives are relatively unconstrained as long as they map to the functional architecture and meet the functional requirements. In general, the greater the design degrees of freedom the greater the potential for influencing performance, life cycle cost and other important measures of design quality. Decisions made on top level architecture during concept design not

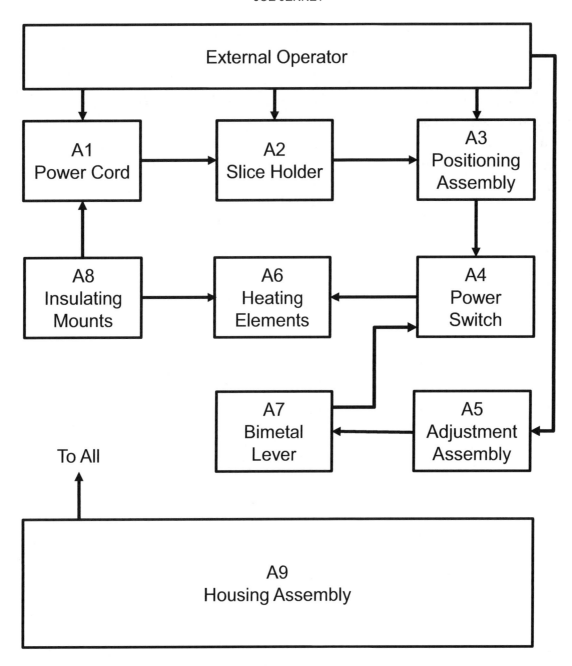

Figure 6-31 A physical block diagram for a candidate toaster concept design.

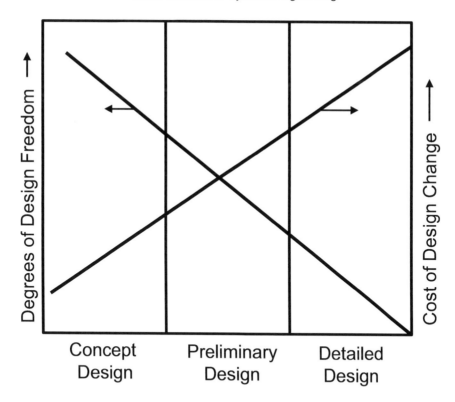

Figure 6-32 Design alternatives cost less to explore and have a greater potential influence during concept design.

only directly reduce the degrees of freedom but these decisions often constrain the design alternatives available at lower levels of the system hierarchy addressed in preliminary and detailed design. This argues strongly for conducting the most extensive exploration of design alternatives during concept design.

The objective is to sufficiently explore alternative concepts to achieve high confidence that the selected design concept is "best" from a number of measures. These measures include the obvious of high performance on high priority customer requirements, low life cycle cost, excellent "ility" measures (manufacturability, testability, repair ability, etc.), and perhaps attractive features that might increase sales. Thus there is a tension between the need to make design decisions quickly and to explore a wide range of design alternatives. Once the desired concept design is established, i.e. a baseline design is defined and trade studies are conducted to select the best alternative for the final baseline, then the design freedom is reduced so the opportunities to significantly improve the design are also reduced.

A standard approach to achieving the desired characteristics in concept designs is to seek **modular designs**. Here the term module refers to design elements, i.e. subsystems, assemblies etc. Modular designs are achieved by refining the allocation of functions to physical modules and partitioning functions between the modules.

6.6.3 Partition for Modular Designs - The DoD SEF says modular designs have the three desirable attributes of **low coupling, high cohesion**, and **low connectivity**. **Coupling** is the amount of information shared between modules; the lower the amount of information that must flow between modules the more independent they are. Having low dependence lowers design risk and makes future upgrades or modifications easier. **Cohesion** is the similarity of tasks performed within a module. High cohesion leads to easier and less complex designs. A design for which a single component performs multiple functions has high cohesion. **Connectivity** is a measure of the internal interfaces between modules. A design that has multiple interconnections between the internal parts of one module and those of a neighboring module has undesirable high connectivity, which again complicates design, integration and testing as well as future upgrades.

Note that **modularity** is a measure of **system complexity**; the higher the modularity the lower the complexity. Risk is a measure of the complexity of the development program; the higher the risk or the more risks that a program has the more complex the development becomes due to the work necessary to mitigate risk. Modularity and risk are related. A system concept design with low modularity is usually higher risk than a design with high modularity. Therefore to achieve the lowest program risk design concepts should be traded to find the highest modularity. However, risk must be evaluated for each concept to ensure that in striving for higher modularity unnecessary risks haven't been introduced.

6.6.3.1 Achieving High Modularity - As the physical design is created many alternatives should be considered. One task is to check that the grouping and sequencing of functions defined during the functional analysis task leads to an effective physical partitioning, i.e. a modular design as described in the previous section. If reasonable physical designs don't result in subsystems being allocated to single functions or single groups of functions then recheck the grouping of functions to see if alternative grouping lead to cleaner physical partitioning and more modularity. It is important to seek clean partitioning of subsystems and their associated functions because the cleaner the partitioning the easier systems are to integrate and test, maintain and upgrade. A function that is implemented in two or more subsystems results in system designs that are more difficult to maintain and upgrade and are often more difficult to test. Envision the design loop as iteration between functional and physical design until both result in a modular physical architecture.

A second task during design synthesis is conducting trade studies, described in a later chapter, to select between design alternatives. Again when evaluating alternative architectures consider the partitioning for each design alternative and examine the possibility that modifying the functional architecture might lead to a better functional to physical allocation and partitioning.

6.6.3.2 Functional to Physical Allocation Matrices - A simple tool that is helpful in refining the functional and physical architectures is a functional to physical allocation matrix. An example matrix for the toaster functional architecture shown in Figure 6-24 and the design concept architecture shown in Figure 6-31 is shown in Figure 6-33. The functional to physical allocation matrix is particularly helpful in examining the partitioning of a design concept for modularity. Typically the more diagonal this matrix the better the modularity. However, opportunities for one physical entity to perform two or more functions are highly desirable and readily apparent in the matrix. Similarly, when the matrix shows a function spread across several physical entities the matrix provides a visual means of examining if the physical design concept is sound or if it should be changed to allow cleaner partitioning. Sometimes the nature of a function causes it to be spread across several physical entities without complicating the design in ways that cause manufacturing, testing or upgrade problems. For example, in Figure 6-32 the "apply heat" function is allocated to three entities and this is probably a reasonable design approach because it likely reduces the parts count and makes operation simpler.

			Functions									
			1.1 Apply Power	1.2 Accept Items for Toasting	1.3 Position Items	1.4 Apply Heat	1.5 Adjust Heating Time	1.6 Time Heating	1.7 Terminate Heating	1.8 Reposition Items	1.9 Prevent Shock	1.10 Provide Structure
Physical Allocation	A1	Power Cord	x									
	A2	Slice Holder		x								
	A3	Positioning Assembly			x	x				x		
	A4	Power Switch				x			x			
	A5	Adjustment Assembly					x					
	A6	Heating Elements				x						
	A7	Bimetal Lever						x	x			
	A8	Insulating Mounts									x	
	A9	Housing Assembly									x	x
		External Operator	x	x	x		x			x	x	

Figure 6-33 A functional to physical allocation matrix for one candidate toaster design concept.

Exercises:

1. Review the "VEE" diagram and the system engineering process diagrams of IEEE, NASA and DoD. Chose the one that you think best explains how systems engineering is done, or should be done, in your organization. Modify the chosen diagram to make it a better fit for your organization.

2. Review the system hierarchy listed in section 6.1. Does your organization follow this hierarchy or another standard hierarchy on every system development?

3. What standard process does your organization use to examine, understand and prioritize the requirements for a new system development provided by your marketing organization or by your customers?

4. Make a list of the documents, diagrams and models your organization generates that belong in the Operational View, Functional View and Physical View of new systems.
5. Identify all of the items listed in question 4 that are candidates for patterns or templates for future system developments.
6. Why should modes be defined for modes other than the operational or in use mode?
7. Is Object Oriented or Process Oriented FFBD better for the systems your organization develops?
8. Why is developing an N-Squared diagram useful for defining interfaces?
9. Name at least six attributes of good requirements statements.
10. How can the specification tree for your systems be organized to facilitate reuse?
11. Why should many alternative system design concepts be developed and traded?
12. What is high modularity and how is it achieved?

7 QUALITY FUNCTION DEPLOYMENT (QFD) IN SYSTEM ENGINEERING

7.0 Introduction

System development is a complex process. Bringing a product or system from concept through production, deployment (distribution) is generally called the **Product Development Process**. Consideration of product development as a process requires looking at a network of tasks that are necessary to bring the product to market and provide a product or system that fulfills the "Voice of the Customer "(VOC). The VOC represents a definition of the needs and wants of the customer. Regardless of how it is presented the product development process is exceedingly complex. It consists of numerous tradeoffs, shared responsibilities and interpretation of differences often resulting in conflicting priorities. A substantial body of technical knowledge must be employed (and deployed) often over a relatively long time frame, while experiencing constant resource changes. The product development process requires a great deal of communication and a substantial work effort from many different functional groups. Product development can no longer be viewed as simply a Design Engineering Function. Design Engineering is but one process of several interrelated processes and functions involved in developing a quality product. Ultimately the communication of information gets back to the customer in the form of a product or service. To minimize the risk and accomplish this effort successfully requires an effective communication and tracking tool and methodology. The implantation of the QFD method can achieve this objective.

A system, however complex, must be carefully planned out to minimize subsequent redesign. The full effect of inadequate planning (or understanding of relationships) is rarely detected until late in the development cycle or not until hardware is fabricated or code written. The later any design change or design defects are detected the more time and money a redesign effort incurs. System Engineering is primarily about understanding relationships and interdependencies during system development. The System Engineering objective is to translate customer requirements to product(s) and service(s) that fulfill the customer's needs. Systems Engineering has emerged as a distinct professional discipline in direct response to the increasing complexity of new development projects for all market applications.

Quality Function Deployment (QFD) is a systematic process for translating customer requirements into appropriate company requirements at each stage from research and product development to engineering and manufacturing to market/sales and distribution. The output of the QFD process is a series of matrices that define critical parameters and requirements throughout the product life cycle. A Product Life Cycle (PLC) is a sequence of stages that a product goes through from conception through design, production, distribution and final phase out, (commonly called cradle to grave)

The prime assumption underpinning system engineering is that a product should be designed to fulfill the customer's actual needs. Self-evident as this approach may seem, it is surprisingly common for companies to develop products with little or no customer input or confirmation of perceived customer needs. Even when a large market exists, a product can fail when the customer's real needs are poorly understood or improperly deployed to the subsequent design process. The QFD process applied to hardware development is very similar and complementary to the recommended System Engineering (SE) process. Much of the information generated by the QFD process is information needed to complete the System and Component specifications. One added benefit that QFD brings to the SE process is a method for prioritization of requirements. OFD also provides a method for identifying the critical design requirements (i.e. Cardinal requirements). QFD provides a pictorial illustration of the System and Component Specifications and shows where requirements are allocated to the subsystems.

The QFD process is complementary and beneficial to decision making in the system development stage. QFD is a structured process to identify these relationships and determine which are most important in driving requirements and system development. QFD generates the skeleton structure (architecture) for the system and subsystem specifications. When QFD is integrated into the system development process, it provides value-added information and knowledge to aid decision making while optimizing the product design

This section:

- Defines QFD and how it can be integrated into the SE and System Development process to provide complementary benefits and aid decision making in defining and specifying a system.
- Explains the benefits and features of using QFD in the System Development process.
- Presents a simple structured system engineering approach to product development using QFD prior to detailed product design.
- Focuses on the hardware development, but in general the concepts can be applied to software development to capture customer needs and requirements and flow requirements through algorithm development

Standard system engineering techniques have been defined in the United States (US) Department of Defense (DoD) and NASA for decades; and these techniques are also applied to large commercial products, e.g. in automotive and aircraft industries. The health care sector has introduced QFD as a means to develop systems for health care. The QFD initiative in the US for hardware developed items follows a structured and disciplined process very analogous to the SE process defined by DoD. DoD SE methodology standardizes the **flow-down and traceability** of specifications for complex products from customer requirements through production, operation, and disposal. SE integrates all of the disciplines and specialty groups into a team effort forming a structured development process that proceeds from concept to production to operation.

The principles of system engineering using QFD span the entire life cycle of a product, but this chapter is concerned with the early feasibility and concept stages. Studies show that a large percent of the product's manufacturing cost is frozen at concept selection time. Many companies in all industries historically initiate product cost reduction efforts after the product is released to the production floor. If say, 85% of the manufacturing cost is frozen at concept development time, then post production release cost reduction is saving cost on 15% of manufacturing effort. The value-added return for the investment is likely to be low. Therefore, to make significant cost reductions, development effort must focus on the early concept selection stage where alternative concepts and technology are considered.

7.1 Background

The word quality in QFD has led to much misunderstanding. QFD was first introduced in most organizations through the Quality Assurance departments. In the QFD process several functional organizations other than the quality department are vital participants. Because the name can be misleading, QFD has been given a bad connotation. QFD is not a quality tool to audit functional organizations, rather it is a structured planning tool to guide and direct the product development process. Let us therefore not be resistant to the use of QFD because of the name, but rather seek to understand what QFD embodies.

Quality Function Deployment (QFD) is a translation of six Japanese Kanji characters:

HIN SHITSU KI NO TEN KAI

As with any translation there is room for other interpretations. Each pair of the Kanji characters has alternate translations; Figure 7-1 illustrates these different translations. The most accepted interpretation is Quality Function Deployment (QFD).

QFD has a broad meaning. It involves taking the features of a product driven by the customer's needs and evolving the product functions into an overall product. We may think of QFD as the act of taking the voice of the customer (VOC) or user all the way through the product development process to the factory floor and out into the market place. QFD is therefore more than a quality tool, but an important planning tool for introducing new products and upgrading existing products.

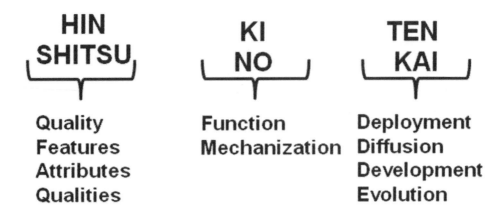

Figure 7-1 The Kanji characters for QFD have several alternate translations.

7.1.1 Features of QFD - As previously stated, QFD:

- Is a systematic means of ensuring that the demands of the customer and the market place are accurately translated into products and/or services.
- Is a structured approach that provides both a planning tool and a process methodology.
- Identifies the most important product characteristics, the necessary control issues and the best tools and techniques to use.
- Applies to all stages of product development and provides a comprehensive tracking tool and communication medium.
- Applies a cross functional team approach combining information and expertise from marketing, sales, design engineering and manufacturing.
- Provides a systematic and disciplined method of creating priorities, making improvements, and defining goals and objectives applicable to the company's products and/or services.

QFD is a method; it is not a panacea, it must be done correctly and it takes up front time and resources to get the best possible results.

7.1.2 Benefits of QFD - QFD is a relatively simple but highly detailed process. Upon initial evaluation it may appear to be too detailed - perhaps not worth the effort. However QFD has proven benefits, including:

1. *A PROPRIETARY KNOWLEDGE BASE;*

    The QFD process leads the participants through a detailed thought process, pictorially documenting their approach. The graphic and integrated thinking that results, leads to the preservation of technical knowledge, minimizing the knowledge loss from retirements or other organizational changes. This use of QFD helps transfer knowledge to new employees, starting them higher on the learning curve. The use of QFD charts results in a large amount of knowledge captured and accumulated in one place. The charts provide an audit trail of the decisions made by the project team. Once a QFD project has been completed, the resulting charts may be used as a starting point for future versions, (a "re-engineering starting point") for similar products. The bottom line of QFD is higher quality, lower cost, and shorter development time resulting in a substantial competitive advantage.

2. **SATISFIED CUSTOMERS**; QFD forces increased understanding of customer requirements because it is driven by the voice of the customer, rather than the voice of the engineer or executive. By focusing on the customer, numerous engineering decisions are guided to favor the customer. Whereas numerous trade-offs are always necessary for any well optimized product, these trade-offs are made for customer satisfaction not for engineering convenience.
3. **FEWER START-UP PROBLEMS**: The preventive approach fostered by QFD results in fewer downstream problems, especially at production startup.
4. **LOWER START-UP COST**; This translates directly into reduced start-up costs
5. **LESS TIME IN DEVELOPMENT**: This approach not only saves money, it also saves overall development time. Product introduction cycle time has been shown to be a third to a half shorter by using QFD to thoroughly plan the product or service.
6. **FEWER FIELD PROBLEMS**; The cost savings has been demonstrated to continue well beyond startup, and is reflected in reduced problems for customers and consequent warranty cost reduction.
7. **FEWER AND EARLIER CHANGES**; A major advantage of QFD is that it promotes preventive rather than reactive development of products. QFD is a preventive approach that has demonstrated fewer downstream production problems; especially at production start-up; commonly referred to as "the transition from development to production".

7.2 The QFD Approach

There are several approaches to QFD; each of these approaches makes use of matrices to organize and relate pieces of data to each other. Many times these matrices are combined to form a basic QFD tool called a **"House of Quality"**.

The basic approach used here is conceptually similar to the practice followed by most American manufacturing companies. In QFD we typically follow the flow as defined in Figure 7-2. We start with customer requirements, which may be loosely stated qualitative items such as: *looks good, easy to use, works well, feels good, safe, comfortable, lasts long, luxurious* or specifically defined requirements. These are important to the customer, but do not represent a product definition. In order to implement a product system engineers need to convert these vague customer requirements into actionable internal company requirements, which we call design requirements. These are generally global product characteristics such that if properly executed the product will satisfy the customer requirements.

Products are not usually developed at this global level, but rather at the system, sub-system or part level. The global design requirements must then be translated into specific product design, company infrastructure and capital investment requirements. Every product has several critical characteristics that determine how good a product fulfills its intended functions. The QFD process allows one to track these critical characteristics throughout the development process. Next determine the required manufacturing operations. This stage is often constrained by previous capital investment. Development organizations usually do not want to build a new factory or install a new line of equipment to produce a new product version and this often constrains product design and production methods.

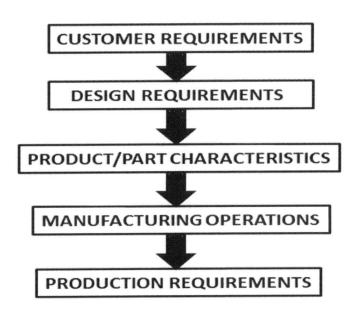

Figure 7-2 The typical flow of applying QFD has five steps.

Within the defined operating constraints, determine which manufacturing operations are most critical to creating the desired critical product and part characteristics, as well as the process parameters of those operations that are most influential. Think of these process parameters as the knobs or dials of the manufacturing operation that are controllable. The manufacturing operations are then deployed into production requirements, which are the entire set of procedures and practices that enable the production system to build products that ultimately satisfy customer requirements.

These operating procedures determine how the factory operates the manufacturing processes to consistently produce the required critical product/part characteristics. They include a number of soft" issues such as inspection and Statistical Process Control (SPC) plans, preventive maintenance programs, operator instructions and training, as well as identifying the need for mistake proofing devices for preventing inadvertent operator errors

The hierarchical approach described above is not unlike the approach taken for years with varying degrees of success. The problem is that some of the translations are not made properly. There are several key reasons for these improper translations that are the result of the structure of large organizations and the complexity of the product development process.

7.3 Hierarchical Matrices and QFD Phases

There are several approaches to the implementation of QFD. The QFD method presented here follows the approach taught by the American Supplier Institute (ASI) based upon the **"House of Quality"** structure. The basic matrix structure consists of various types of matrix and table sections (or "rooms") linked together to form what has been termed the "House of Quality". The ASI approach implements a four phase approach with matrices representing the important characteristics for each phase. Figure 7-3 illustrates the four phases and how the key characteristics of each phase are deployed to the next phase of development.

The four phase approach results in a hierarchical series of matrices where each individual matrix is called a quality table (QT) and are numbered, e.g. QT-1 thru QT-4. In the four phase approach a team determines the relationships between customer requirements and product design requirements. The product design requirements are then deployed to the QT-2 matrix. In QT-2, the team determines the relationships between design requirements and product design. The product design is then deployed to the QT-3 matrix to determine the relationships between product design and process design. The process design is then deployed to QT-4 matrix to determine the relationship between process design and manufacturing operations.

The four phase hierarchical series of matrices when completed links the customer requirements all the way to the manufacturing operations. Thus as the manufacturing operations meet their deployed requirements then indirectly the customer requirements are being satisfied. Therefore the resulting product or system is Customer Driven.

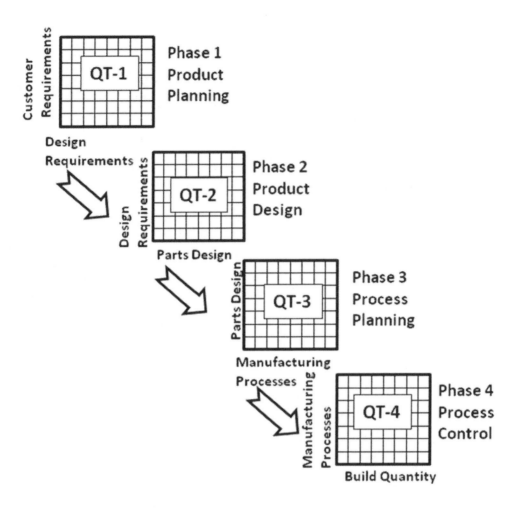

Figure 7-3 The four phases of QFD consist of four linked matrices called Quality Tables.

7.4 Basic Matrix Structure

The QFD process is represented by a series of interconnecting matrices that establish the WHAT, the HOW's and the interrelationship of all parameters involved in the product development process. The QFD method is simply a disciplined way of deploying the voice of the customer (VOC) through each stage of the product cycle. The objective is to keep all efforts focused on the VOC requirements, to optimize cost and to minimize cycle time while being driven by the VOC.

The QT's are used in each product phase to communicate the knowledge developed to the next stage. In each stage translations take place to systematically bring the VOC to actions taken by functional organizations that result in a product/service that satisfies the customer. The purpose of the QT charts is to focus on answering three questions; WHAT, HOW and HOW MUCH. For each product stage and for each action taken in that stage these three questions must be addressed.

The "House of Quality", sometimes referred to as the *Enhanced House of Quality* consists of multiple "rooms". Four of the "rooms" are lists that capture the, "What's, How's, Why's and How Much's" of the project. Four additional "rooms" are formed by determining the correlation and relationships between these lists. Figure 7-4 illustrates the basic structure and location of these "rooms". The following sections provide detail in forming the lists and relationships between these "rooms" that make up the "House of Quality". All four phases of the hierarchical matrices follow this basic structure and form.

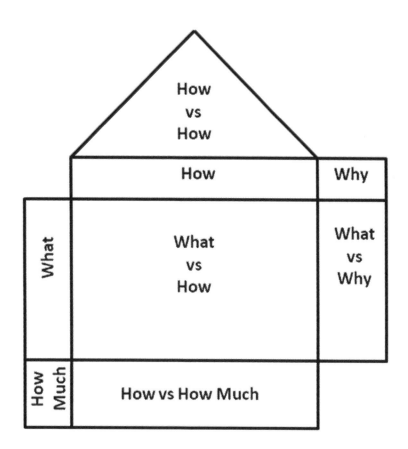

Figure 7-4 The rooms and relationships of the house of quality.

7.4.1 Voice of the Customer (The "What's") - QFD starts with a list of objectives, or the WHATs that we want to accomplish. In the context of developing a new product this is a list of customer requirements and is often called the Voice of the Customer (VOC). The items contained in this list are usually very general, vague and difficult to implement directly; they require further detailed definition. These vague needs are sometimes called "verbatims", (e.g. easy to use, lasts long time, light weight, low power, easy to modify).

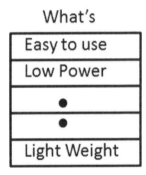

Figure 7-5 The "what's" defined by the VOC are often general statements.

One such item might be "easy to test", which has a wide variety of meanings to different people. This is a highly desirable product feature, but is not directly actionable.

7.4.2 Transformation of Action - Once the list of WHAT's is developed, each requires further definition. The list is refined into the next level of detail by listing one or more HOW's for each WHAT, (i.e. How are we going to satisfy the WHAT's) as shown in Figure 7-6. This process can be further refined and expanded into a more detailed list of HOW's.

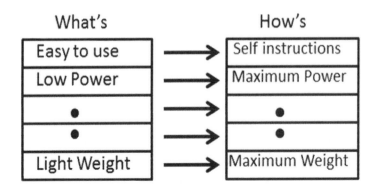

Figure 7-6 The list of WHAT's are transformed into a list of HOW's

The objective of this refinement is to identify each actionable requirement - one that a clear action taken will satisfy a WHAT.

7.4.3 Handling Complex Relationships - The problem that is encountered is depicted in Figure 7-7. Many of the HOWs identified affect more than one WHAT. The approach to charting the `WHATs and `HOWs sequentially would become a maze of lines due to interrelationship that exist between the parameters

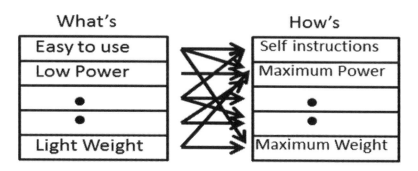

Figure 7-7 Many HOW's affect more than one WHAT

7.4.4 Structuring the Relationships in a Matrix - The complexity of the sequential process is solved by creating a matrix with the HOW list across the top (horizontally) and the WHAT list vertical down the side of the matrix. This determines the RELATIONSHIPS of the WHAT's and HOW's in a matrix where each intersect. This is called a RELATIONSHIP MATRIX. Figure 7-8 illustrates by the use of a "X" where the What's and How's are interrelated.

	Self instructions	Maximum Power	•	•	Maximum Weight
Easy to use	X	X			X
Low Power		X			
•		X		X	X
•			X	X	X
Light Weight		X			X

Figure 7-8 A correlation matrix determines the relationships between the WHAT's and HOW's

7.4.5 Kinds of Relationships - The RELATIONSHIPS are the third key element of any QFD matrix and are depicted by placing symbols at the intersections of the WHATs and HOWs that are related. It is possible to depict the strength of the relationships by using different symbols. Commonly used symbols are shown in Figure 7-9.

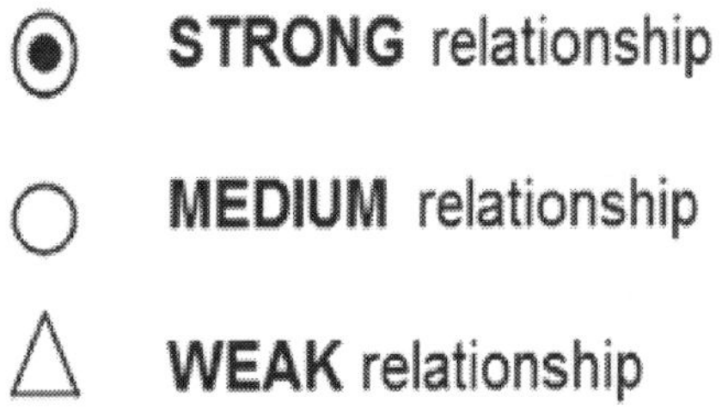

● **STRONG** relationship

○ **MEDIUM** relationship

△ **WEAK** relationship

Figure 7-9 Symbols used to show the strength of relationships.

This method allows very complex relationships to be depicted graphically and is easily interpreted as shown in Figure 7-10.

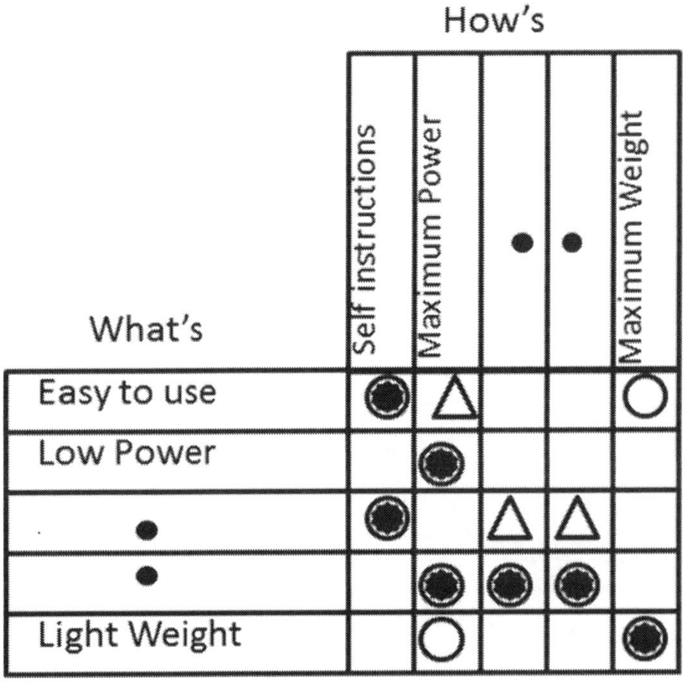

Figure 7-10 Strength symbols are placed in the matrix relating each WHAT to its respective HOW's.

Throughout the QFD process there are repeatedly opportunities to cross check thinking, thus leading to better and more complete designs. This technique of evolving plans into actions is useful for new product development as well as applications in business planning and systems design.

7.4.6 Target Values (How Much) - The fourth key element of any QFD chart is the HOW MUCH section. These are the measurements for the HOWs. These target values should represent what is necessary to satisfy the customer and may not be current performance levels. "Easy to test", when translated into detailed requirements may be measured in terms of the **number of test points**, requirement for **component spacing,** component **edge clearance,** etc. The component clearance would be a

HOW and the HOW MUCH would be 0.020 inches minimum. HOW MUCH's are needed for two reasons:

1. To provide an objective means of assuring that requirements have been met.
2. To provide targets for further detailed development

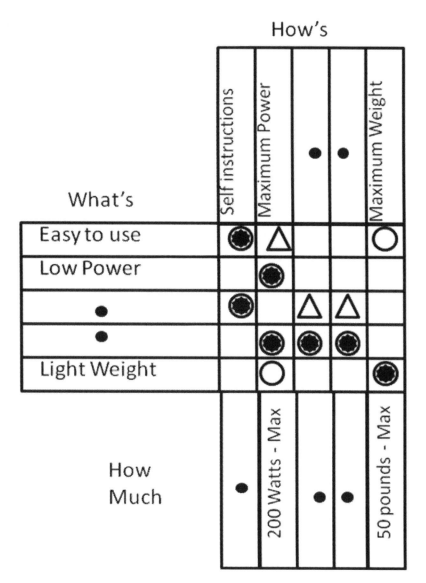

Figure 7-11 The HOW MUCH's are added in rows at the bottom of the matrix.

The HOW MUCH's, added to the matrix as shown in Figure 7-11, provide specific objectives that guide the subsequent design and afford a means of objectively assessing progress, minimizing **opinion-eering**". The HOW MUCH's should be measurable as much as possible, because measurable items provide more opportunity for analysis and optimization than do non-measurable items. This aspect provides another cross check on thinking. If most of the HOW MUCH's are not measurable then the definition of the HOW's are not detailed enough. The HOW relationships that relate to the WHAT's become one means to check and measure to see if the WHAT requirements are being met. Viewed another way; meeting the target values will satisfy the HOW requirement. If all of the HOW requirements are satisfied that are related to a VOC item by the relationship matrix then the VOC item is met. Therefore the focus can be now on meeting the target values and not be directly concerned with the VOC, it is taken care of by fulfillment of the HOW MUCH's. These four key elements (WHAT, HOW, RELATIONSHIPS, HOW MUCH) form the foundation of QFD, and can be found on any QFD chart.

7.4.7 Correlation Matrix - The CORRELATION MATRIX is a triangular table often attached to the HOWs, establishing the correlation between each HOW item. The purpose of this roof-like structure is to identify areas where trade-off decisions, conflicts and research and development may be required. As in the RELATIONSHIP MATRIX, symbols are used to describe the strength of the relationships. The CORRELATION MATRIX also describes the type of relationship. The symbols commonly used are shown in Figure 7-12.

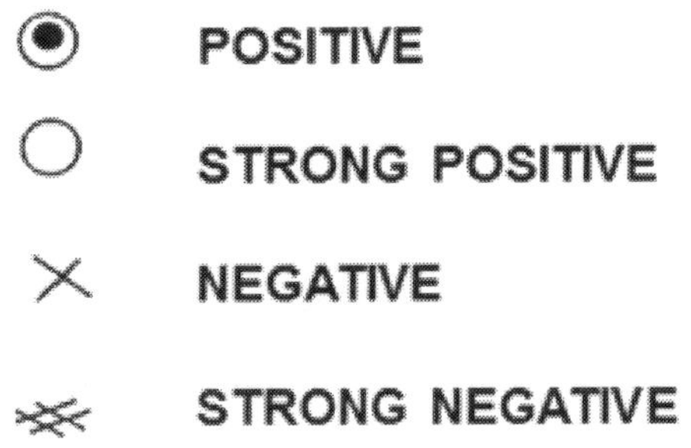

- ⦿   **POSITIVE**
- ○   **STRONG POSITIVE**
- ✕   **NEGATIVE**
- ⚹   **STRONG NEGATIVE**

Figure 7-12 Symbols used to indicate correlation between pairs of HOW's.

The correlation matrix identifies which of the HOWs support one another and which are in conflict. Positive correlations are those in which one HOW supports another HOW. These are important because some resource efficiencies are gained by not duplicating efforts to attain same result. If an action adversely affects one HOW, it will have a degrading effect on the other. Negative correlations are those in which one HOW adversely affects the achievement of another how. These conflicts are extremely important; they represent conditions in which trade-offs are suggested. If there are no negative correlations there is probably an error. A well optimized product is almost always the result of some level of trade-off, which is expressed by a negative correlation.

Generally every HOW MUCH item has a desired direction. For example, POWER of 100 watts; generally driving it lower is better. A good test for determining if a relationship is positive or negative is to ask the question: "If power is driven towards its desired direction, are the other HOW's driven toward or away from their desired target values? If the HOW is driven towards its desired target value when power goes towards its desired target value then it is a POSITIVE RELATIONSHIP. If it is driven away from its desired target value then it is a NEGATIVE RELATIONSHIP."

Be cautious not to jump to a trade-off too quickly. The goal is to accomplish all of the HOW's in order to satisfy customer requirements. The response to a negative correlation should be to seek a way to make the trade-off go away. This may require some degree of innovation or a research and development effort that may lead to a significant competitive advantage.

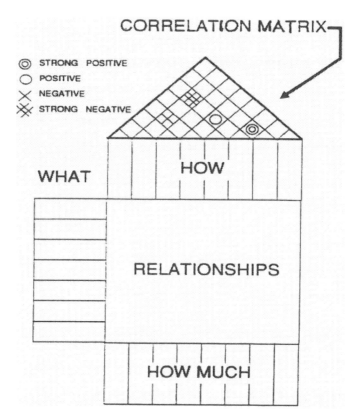

Figure 7-13 The correlation matrix is constructed on top of the HOW's.

Frequently, negative correlations indicate conditions in which design and physics are in conflict. When this occurs physics always wins. Such trade-offs must be resolved. Trade-offs that are not identified and resolved often lead to unfulfilled requirements even though everyone has done their best. Some of the trade-offs may require high level decisions because they cross engineering group, department, divisional or company lines. Early resolution of these trade-offs is essential to shorten program timing and avoid nonproductive internal iterations while seeking a nonexistent solution.

Trade-off resolution is accomplished by adjusting the values of HOW MUCH's. These decisions are based on all the information normally available; business and engineering judgment as well as various analysis techniques. If trade-offs are to be made, they should be made in favor of the customer and not what is easiest for the company to perform.

7.4.8 Competitive Assessment - The COMPETITIVE ASSESSMENT is a pair of graphs that depict item for item how competitive products compare with current company products. This is done for the WHAT's as well as the HOW's. The COMPETITIVE ASSESSMENT of the WHAT's is often called a Customer Competitive Assessment, and should utilize customer oriented information. It is extremely important to understand the customer's perception of a product relative to its competition.

The COMPETITIVE ASSESSMENT of the HOW's is often called a **Technical Competitive Assessment**, and should utilize the best engineering talent to analyze competitive products. The COMPETITIVE ASSESSMENT can be useful in establishing the value of the objectives (HOW MUCH's) to be achieved. This is done by selecting values which are competitive for each of the most important issues. The COMPETITIVE ASSESSMENT provides yet another way to cross check thinking and uncover gaps in engineering judgment. If the HOW's are properly evolved from the WHAT's, the COMPETITIVE ASSESSMENTs should be reasonably consistent.

WHAT and HOW items that are strongly related should also exhibit a relationship in the COMPETITIVE ASSESSMENT. For example, if we believe superior dampening will result in an improved ride, the COMPETITIVE ASSESSMENT would be expected to show that products with superior dampening also have superior ride; as illustrated in Figure 7-14.

If this does not occur, it calls attention to the possibility that something significant may have been overlooked. If not acted upon, we may achieve superior performance to our "in house" tests and standards, but fail to achieve expected results in the hand of our customers.

The IMPORTANCE RATING is useful for prioritizing efforts and making trade-off decisions. Numerical tables or graphs will depict the relative importance of each WHAT or HOW to the desired end result. The WHAT IMPORTANCE RATING is established based on customer assessment. It is expressed as a relative scale (typically 1-5) with the higher numbers indicating greater importance to the customer. The importance ratings are listed in a column between the WHAT's and the matrix. It is important that these values truly represent the customer, rather than internal company beliefs. Since we can only act from the HOW's, importance ratings for these HOW's are needed.

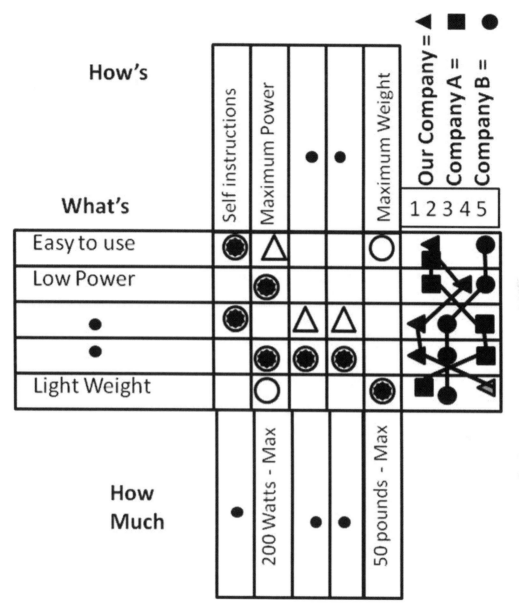

Figure 7-14 Competitive assessment of the WHAT's are put in a box on the right side of the matrix.

7.4.9 Importance Ratings - Weights are assigned to the RELATIONSHIP symbols, e.g. the 9-3-1 weighting shown in Figure 7-15 achieves a good variance between important and less important items. Other weighting system may be used. For each column (or HOW), the WHAT importance value is multiplied by the symbol weight, producing a value for each RELATIONSHIP. Summing these values vertically defines the HOW importance value. In Figure 7-16 the HOW importance rating for the first column is calculated in the following manner. The double circle symbol weight (9) is multiplied by the WHAT importance value (5), forming a RELATIONSHIP value of 45. The next double circle symbol weight (9) is multiplied by the WHAT importance value (2), forming a RELATIONSHIP value of 18. These two values (45 + 18) form the HOW importance value of 63. This process is repeated for each column as shown in Figure 7-16.

STRONG = 9
MEDIUM = 3
WEAK = 1

Figure 7-15 Importance ratings are obtained by assigning weights to the symbols in the relationship matrix

The IMPORTANCE RATING for the HOW's provides a relative importance of each HOW in achieving the collective WHAT's. We see that for the HOW's listed; "Maximum Power" with a Target Value of 200 watts has the "HIGHEST" relative importance. Greater emphasis should be placed on the HOW with the 83 rating than the other HOW's. It is important that we are not blindly driven by these numbers. The numbers are intended to help us, not constrain us. Look upon the numbers as further opportunities to cross check thinking. Question the relative values of the numbers in light of judgment. Is it reasonable that the HOW valued at 83 is the most important? Is it reasonable that the HOW's with similar ratings are nearly equal in importance?

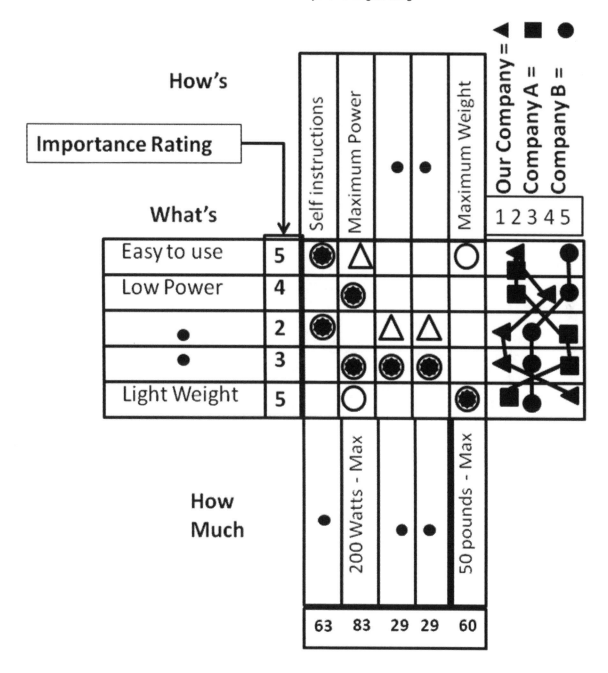

Figure 7-16 Importance ratings are calculated for each HOW as the sum of the weighted importance of each WHAT.

7.4.10 The Basic Matrix Structure - The previous section can now be integrated together into one chart. Figure 7-17 illustrates the Basic Matrix Structure. All of the matrices used in the product development stages could have these basic sections. Note the correlation matrix when added to the relationship matrix takes on the shape of a house with a roof. It is from this construction that the QFD matrices are termed the "houses of Quality".

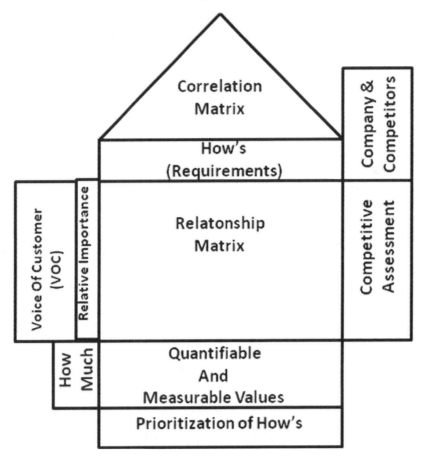

Figure 7-17 The basic form of the House of Quality relates the VOC and competitive assessment information to design requirements.

7.5 Other Specialty Matrices

As previously stated, there are several approaches to QFD; each of these approaches makes use of matrices to organize and relate pieces of data to each other. There are typically four basic QFD matrices for Product Development. The basic structure of these matrices relates WHAT's to HOW's for each of the four product development stages. Since one of the objectives for System Engineering is to understand and determine relationships at all stages, there are other WHAT vs. HOW relationships that are important. Two "Specialty Matrices" that are important in front end system development are: 1) *Preplanning Matrix* and 2) *Functional Architecture* of the system (the WHAT's) versus the structure of the Product Design, (the HOW's)

7.5.1 Pre-Planning Matrix (PPM) - A Pre-Planning Matrix (PPM) is sometimes used in the product planning phase prior to launching Product Development. The PPM provides an assessment of current company capabilities versus the expected competition. The assessment is taken from where you believe the Customer rates your capabilities and your competition. The assessment identifies current strengths and weakness and shows where investment is needed to create a better competitive position prior to entering into the pursuit and/or development of a product. It also provides a method of identifying potential teaming, strategic partnerships and/or alliances to strengthen product development and improve competitive position. The PPM provides a means for prioritizing investments that might be needed to become competitive and increase the probability of win or sale.

The PPM also identifies potential **Sales Points**. A Sale Point is where current capabilities or the combined capabilities of a team provide a leading competitive position. The Sales Point is used to communicate with customer(s) and to emphasize why a team or product is superior to its competitors.

The PPM consists of the following sections; 1) Voice of Customer (VOC) requirements, 2) An order of importance of VOC, 3) Competitive Assessment, 4) Identification of probable Sales Points,.5) Improvement Factor of VOC, 6) A new weighted VOC,

Figure 7-18 illustrates a partial sample of a PPM. In reviewing the matrix, Section A in Figure 7-18 lists each VOC requirements followed by a column that rates importance level of VOC on a scale of 1 to 5, with 1 rating being least important and a 5 rating being most important. Section B in Figure 7-18 provides an assessment of the competition and your current capability through the eyes of your customer for your company and each competitor, followed by an identification of probable Sales Points.

The PPM as constructed so far shows that our present competitive assessment indicates that Company B has a better competitive position. The PPM also indicates that Our Company presently has two Sales Points to emphasize; that our product is *Easier to Use* and *Easier to Hold* versus competitor Company A and Company B.

Since the PPM from Figure 7-18 indicates that Company B has a better expected competitive position than Our Company for the defined VOC and Importance Rating, something must be done by Our Company to better the competitive position. Figure 7-19 illustrates a further expansion of the PPM to include additional sections. Section C includes Our Company Plan to increase capability in the "eyes of the Customer". The column marked Our Target Position identifies the improvements needed to increase the probability of convincing the Customer that Our Company has a better capability and therefore provides a more favorable chance of winning the Customer's approval. However to achieve that Target Position Our Company needs to improve in two areas of the Customer's view of our capabilities; 1) Waterproof and 2) Light Weight. Light Weight has an Importance Rating of 5 and it requires an improvement of 9:1 from our current position, Our Company gives this a top priority action, (Priority #1). The final column provides for a summary of Our Company's Strategy to go forward to capture the customer's approval. Accomplishing this development and partnering effort should provide Our Company with a strong competitive position over Company B, (i.e. a Relative Importance Rating of 162 versus Company B's 101 rating).

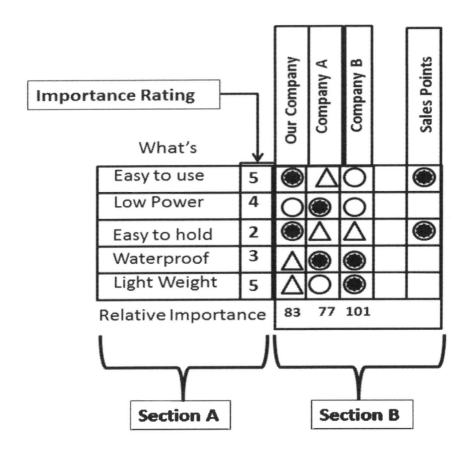

Figure 7-18 The initial PPM is constructed to show current competitive position and identify Sales Points.

The PPM as presented assumes that Company A and Company B do nothing to improve their competitive position, but in reality that is not likely to be the case. A further extension of the PPM is to forecast where Our Company thinks Company A and Company B might make improvements. For example if Company A and Company B were to establish a Strategic Alliance and combine their capabilities then the best of each company capability would result in a Relative Importance Rating of 125, which is still lower than our anticipated improved rating of 162. This assumes that Our Company is able to implement the identified strategy and investments and convenience the Customer of its improved capabilities.

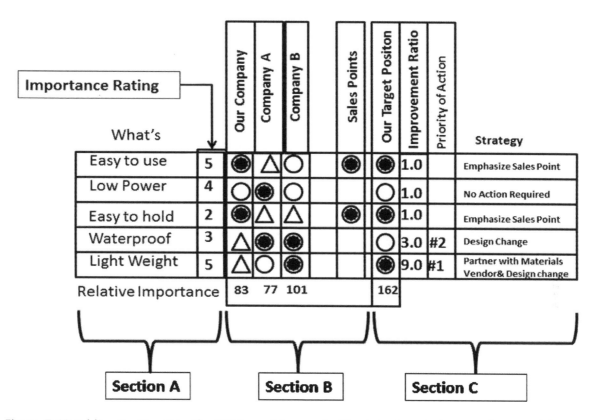

Figure 7-19 Adding Section C to the PPM provides a prioritized strategy to improved competitive position.

There are other methods for the development of the PPM instead of the non-linear ratings used in Figure 7-19. For example instead of the non-linear symbols a numerical rating of 1 to 5 can be used to assess Our Company and Company A and B's capability. However the overall benefit of the PPM is the process of walking through the questions and assessment needed to establish a company's competitive position and provide priority and direction for further investment in development and/or teaming with another company.

It must be noted that other than identifying and emphasizing current Sales Points, nothing is attributed to the Relative Importance Rating for having these Sales Points over the competition. Some PPM methods assign a value of 1.5 for each Sales Point and multiply the VOC item by 1.5. If that had been done in this example the first Relative Importance Rating for Our Company would be 114.5 versus Company A's 77 and Company B's 101. This approach would influence whether Our Company might alter its strategy on investment or teaming. The decision would depend on Our Company's ability to convince the Customer that these Sales Points provide increased value. It is easy to error based on self-evaluation and could lead Our Company into a false sense of competitive position.

7.5.2 Function vs. Product Design - One of the basic building blocks of QFD is the identification of the functions that a product or service must provide. Every product or service has a basic all-encompassing purpose. The primary or basic function is identified as the prime reason for the product's or service's existence. Functional Analysis/Allocation (FA/A) is an early step in the system engineering process, that defines a baseline of functions and sub functions and an allocation of decomposed performance requirements. The FA/A task is to create a functional architecture that provides the foundation for defining the system physical architecture through the allocation of function and sub function to hardware-software and/or operations (i.e. personnel).

It should be clearly understood that the term "functional architecture" only describes the hierarchy of decomposed functions and their allocations of performance requirements to functions within the system. It does not describe either the hardware architecture or the software architecture of the system. It describes "what" the system will do, not how it will do it. Therefore once the functional architecture of the system is defined and a conceptual baseline approach is generated the correlation of functional architecture to physical (hardware & software) architecture needs to be generated to determine:

- Where the function and sub functions are going to be performed in the physical design of the system and
- If all of the functions and sub functions are being performed
- The interfaces and what must cross each interface to perform the intended functions.

Figure 7-20 illustrates an example of a QFD matrix that correlates Functional Architecture with one potential Physical Architecture of the system. This QFD matrix provides a graphical picture of the coupling of Functional Architecture (What is to be accomplished) and Physical Architecture (How and where function is accomplished) along with were interfaces are required. Customers would likely define the primary function or all-encompassing purpose of a camera as "Take a Picture". The systems engineer decomposes this primary function into the needed functions of *Capture an image of a scene, Store the image, Display the image and Readout the image.* The example illustrated in Figure 7-20 relates the needed function *Capture an Image of a scene* and the relationship of this function and its sub functions to a potential hardware concept.

In principle functions and sub functions can be defined totally independent of the technologies used in implementing the functions. However, often a decision is made on one or more of the technologies to be used as the result of market analysis or constructing a PPM. To complete this decision may require Technology Trade Studies to select the best technology approach. Let's assume a choice between a photographic process using a light sensitive material and a digital technique using a charge coupled sensor where the image is electronically captured and stored on a focal plane of electronic detectors. Assume the decision after making a trade study of the advantages and disadvantages of both technologies is to capture the image electronically using an electronic device known as Charge Coupled Device (CCD). The sub functions that are associated with an electronic method of image capture include:

- Convert image light to electrical signal

- Focus light
- Control light intensity (to match CCD detector sensitivity)
- Condition light
- Support & protect components
- Block unwanted light
- Position light
- Support & protect components

		CCD focal plane & electronic signal processing	Entrance focus lens & housing	Baffled Entrance aperture	Mechanical housing
Capture Image	Convert image light to electrical signal	●			
	Focus light		●		
	Control light		●	●	
	Condition light		●		
	Block unwanted light		●	●	●
	Position light		●		
	Support & protect components				●

Figure 7-20 An example functional to physical matrix for the Capture Image function of a digital camera.

A simplified baseline concept is to implement an electronic sensing material called a Charge Coupled Device (CCD). The subsystem components that make up the hardware tree for a camera using an electronic sensing approach are:

- CCD focal plane and electronic signal processing
- Entrance focus lens that places the image at the CCD focal plane
- Mechanical housing that proved structural rigidity and blocks outside light from CCD focal plane
- Entrance aperture baffled to block stray light

Ideally a function is performed in only one module or subsystem. This simplifies the interfaces, however this may be impossible to achieve. In reality functions are likely to be accomplished over more than one subsystem and therefore over one or more interfaces. Figure 7-20 provides a summary of a Functional to Physical QFD Matrix for this example and illustrates the following information:

- All but two sub functions are achieved in an individual subsystem
- Two sub functions, Control Light and Block unwanted light are accomplished in more than one subsystem. Therefore any performance requirements associated with controlling light and blocking unwanted light are allocated over more than one subsystem
- The subsystem Entrance focus lens & housing must support more than one function. In this case it supports five different sub functions.

When more than one sub function is accomplished in the same physical subsystem there can be potential interaction or interdependence of sub functions. Also as the number of sub functions increases in any one subsystem complexity also increases. In this example an analysis should be performed to determine if functional performance of the sub functions interacts or is influenced by the other functions in the subsystem. Tradeoff of performance may be necessary to balance out overall performance. These considerations are not readily apparent without construction of the Functional to Physical QFD Matrix.

It has been documented in Product Development activities that the Functional Architecture for a product or service varies little from one product version to another of the same family. The Functional Architecture is not an invariant, but will change only slightly from one product version to the next. The application of different technologies and design changes cause the hardware/software tree to be different from one product version to the next. The Functional Architecture can be considered as a "re-engineering point" from one product release to the next. Therefore the Functional Architecture can be considered as a pattern of functions to be either reused or to be a starting point for new product/service development.

7.6 Summary

Quality Function Deployment (QFD) is a systematic process for translating customer requirements into appropriate company requirements at each stage from research and product development to engineering and manufacturing to market/sales and distribution. QFD is a complement to the System Engineering process. The principles of system engineering using QFD span the entire life cycle of a product. QFD is not a quality tool to audit functional organizations, but is a structured planning tool to guide and direct the product development process.

In summary QFD:

- Is a systematic means of ensuring that the demands of the customer and the market place are accurately translated into products and/or services.

- Provides both a planning tool and a process methodology in a structured approach.

- Identifies the most important product characteristics.

- Provides a comprehensive tracking tool and communication medium when applied to all stages of product development.

- Applies a cross functional team approach combining information and expertise from marketing, sales, design engineering and manufacturing.

The QFD process leads the participants through a detailed thought process, pictorially documenting their work. The graphic and integrated thinking that results leads to the preservation of technical knowledge; minimizing the knowledge loss from retirements or other organizational changes. This use of QFD helps transfer knowledge to new employees, starting them higher on the learning curve. The use of QFD charts results in a large amount of knowledge captured and accumulated in one place. The charts provide an audit trail of the decisions made by the project team. Once a QFD project has been completed, the resulting charts may be used as a starting point for future versions, (a "re-engineering starting point") for similar products.

Again, QFD is a method; it is not a panacea, it must be done correctly and it takes up front time and resources to get the best possible results.

Exercises:

There are no defined exercises for this chapter. If this material is being studies by a team then it is suggested that the team develop a QT-1 table for a product their organization is developing.

8 SELECTING THE PREFERRED DESIGN

8.0 Review and Introduction

Selecting the preferred design involves several types of systems analysis tasks including: trade studies, risk assessment, cost modeling, performance analysis/modeling and simulation. Before describing some of these processes it is helpful to review several points from chapters 5 and 6. Trade studies, assessment and analysis are involved in each of the three major steps in the systems engineering process. Referring back to the systems engineering process described in Figure 6-4 shows that iteration takes place not just between each of the major steps of requirements analysis, functional analysis and design synthesis but also between each of these steps and the systems analysis processes of trade studies, modeling and simulation included under technical management. Figure 5-1 illustrates how these systems analysis tasks support the decisions involved in flowing down requirements. Chapter 6 discussed how functional analysis is involved in the flow down of requirements and how systems analysis supports allocating requirements to functions. Chapter 6 also discussed the value in considering many design alternatives at each stage of design synthesis but particularly during concept design. The preferred process for considering design alternatives is to arrive at a baseline design and then conduct trade studies of alternative designs. It is systems analysis processes that aid in selecting the preferred design from the alternatives studied. This chapter describes methods for conducting trade studies, provides an overview of modeling and simulation and introduces some additional diagrams that are useful in describing design concepts.

8.1 Baseline Concept Management

Project leaders must strive for a balance between forcing decisions too early so that promising alternatives are not considered and leaving too many decisions until just before major design reviews so that the design details are unclear. The standard process for achieving this balance is forcing a **design baseline** as soon as possible and then conducting trade studies of alternatives to the baseline. This facilitates having a controlled process for making decisions to adopt changes to the baseline and it enables all team members to have the same view of the baseline design at any time.

As soon as the top level function and physical diagrams, the system level specification document and the ICD are complete declare this documentation to be the **baseline design**. It's ok that many assemblies or even subsystems are immature; just use the team's best guesses, but do include as much detail as is available. The baseline design documentation should be maintained where it is available to all team members but no changes to the documentation are allowed without a formal decision process being executed.

As the design progresses and trade studies are performed use the baseline as the starting point and the reference for trades and design decisions. If analysis or trade studies suggest the baseline should change and the project leadership, including the lead systems engineer, agrees then the change is made and the lead systems engineer should notify all key engineering participants that a design change to the baseline has been made. Then the baseline documentation should be updated.

Following this simple baseline management process ensures the whole team is working from the same design data base at any time and it facilitates a controlled design decision process.

8.2 Trade Study Methodology

It is a good practice for a systems engineering organization to adopt a standard methodology for conducting trade studies. A standard methodology makes control and management easier and the discipline of following a standard methodology usually results in faster and better trade results and gives management more confidence in the results. Independent of the methodology used trades must be defined and planned. Key top level trades should be identified in the SEMP but not all trades can be listed in the SEMP. Trade Tree diagrams are useful for defining trades if there are only a few alternatives and a few levels under consideration. **Trade Tree diagrams** are decision trees without the chance values applied to each node. Trade Trees and Decision Trees are useful tools well defined in the NASA *Systems Engineering Handbook*. If there are many trades to be conducted for a single level then **N-squared diagrams** are more useful than Trade Trees. The value of the N-squared diagram is that it provides an easy way to determine the best order to conduct trades and conducting the trades in the best order saves time and money. An N-squared diagram for trades is developed following the same process as for defining and ordering tasks using N-squared diagrams as described in Chapter 4.

One standard trade study methodology is presented here and it is defined for physical design trades. Experience shows that this methodology is useful for design trades at all levels of a system hierarchy. This methodology combines **Design Trade Matrices** with **Pugh Concept Selection** and is best described by the diagram shown in Figure 8-1.

Brainstorming is a good way to identify alternatives and many alternatives should be defined. It is important to select people with a diversity of skills and experience for effective brainstorming. Otherwise the alternatives may not have the variety and innovativeness desired. If a large number of alternatives is identified the number of alternatives can be thinned to narrow the trade space. This is accomplished by point design analysis or by using engineering judgment to select a subset of the "best" concepts. The next step is to define selection criteria. These criteria are derived from the requirements. If a process like QFD is used there are a set of requirements that are more important to customers than others. These are often called **cardinal requirements** or key requirements and they are the ones most likely to be the important criteria that should be used in trade studies.

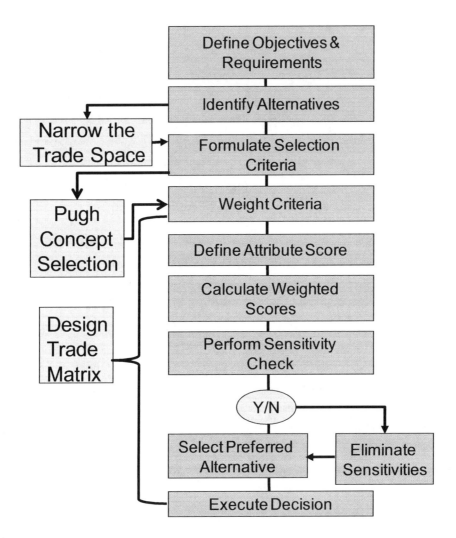

Figure 8-1 An excellent trade study process combines Pugh Concept Selection with a Design Trade Matrix.

8.2.1 Pugh Concept Selection - Having defined selection criteria the next step is to refine the concepts. Pugh Concept Selection is the critical step that leads to improved concepts. If large or complex system designs are being traded then first conduct Pugh Concept Selection at a sub system level. Then combine the best results into a top level system design concept.

Pugh Concept Selection is conducted by choosing one alternative concept as the benchmark. Each of the other alternative concepts is then compared to the benchmark. The objective is to use the results of the comparison to suggest new concepts or even new criteria. A matrix diagram is used to compare the alternative concepts to the benchmark as shown in Figure 8-2.

Criteria	Concept a	Concept b	Concept c	Benchmark
1	+	-	-	■
2	+	S	-	■
3	-	+	+	■
Score:	+1	0	-1	■

Figure 8-2 A Pugh Concept Selection matrix for three alternative concepts compared to a benchmark concept.

For each criteria each concept is compared to the benchmark concept and a decision is made as to whether the concept is better (+1), worse (-1) or the same (S) as the benchmark. Weighting should not be used at this point in the trades. The score for each concept is obtained by summing the number of pluses and minuses, each counting as a plus one or minus one and the S counting as zero. The next step is to examine the concept that scores the best, assuming that one scores better than the benchmark. For example, in Figure 8-2 concept a scores better than the benchmark and the other two concepts. Note however that for criteria 3 concepts b and c are superior to both the benchmark and concept a. Examine the reasons for concepts b and c being better for criteria 3 and see if changes can be made to either concept a or to the benchmark that will turn concept a's minus score for criteria 3 to a plus or a same. The idea is to consider new alternatives that combine the best features of the traded concepts to arrive at a better concept. Thus the objective of Pugh Concept Selection is not to select the best of the alternatives but to define new concepts that are better than any of the initial concepts.

8.2.2 Design Trade Matrices - The Pugh Concept Selection process can be repeated until no new concepts are suggested that are better than the existing set. When a final set is agreed upon the next step is to conduct a weighted trade using a **design trade matrix** (decision matrix in NASA nomenclature). A design trade matrix might look like Figure 8-3 for three candidate concepts.

Again if QFD is used the weights for the criteria can be drawn from the QFD analysis. If not then engineering judgment can be used for the weights. In the example shown the weights are selected over a range from 1 to 5. An alternative is to use percentages that add to 100 percent for all criteria. Attribute scores can be 1, 2 or 3 or 1, 3 or 9 or the actual results of using an analytical tool. If different tools are used for different criteria the attribute scores can be normalized to a fixed range for all criteria to maintain the validity of the selected weights.

Criteria	Weight	Evaluation			Score		
		Concept A	Concept B	Concept C	Concept A	Concept B	Concept C
1	3	1	2	3	3	6	9
2	5	3	3	2	15	15	10
3	2	2	3	1	4	6	2
				Total Score	22	27	21

Figure 8-3 An example design trade matrix for three concepts and three weighted criteria.

The step following determining the total scores is to perform a sensitivity check so ensure the results are significant. One method of performing a sensitivity check is to examine evaluation values for criteria with large weights. If a small change in one evaluation changes the total score to favor a different concept that the results aren't reliable. If the results are not reliable then consider adjusting the weights, adding additional criteria or using a more sensitive method of determining attribute scores.

8.2.3 Pitfalls for Trade Studies – Common mistakes that can lead to ineffective trade study results include:

1. Poor requirements definition can result in a trade result that may not be good for properly defined requirements.
2. Valuable alternatives may be missing if alternatives are defined without brainstorming by several experienced people with a diversity of skills and experience.
3. Allowing biased weightings or selection criteria often results in selecting alternatives that are driven by the biases and not the optimum alternative that would be selected with unbiased trades.
4. The fatal error is having no winner. This results if the spread of the weighted score is less than the spread of estimates of errors. The sensitivity analysis step in Figure 7.1 is crucial to effective trades.
5. Inappropriate models used for determining attribute scores. Models not only have to be relevant to the trade being performed they must have credibility in the eyes of the decision makers, they should lead to scores for the different alternatives that are spread more than the estimates of errors in the model results, the algorithms and internal mathematics must be transparent to the users and they must be sufficiently user friendly that the analysis can be conducted with confidence and in a timely manner.

6. Conducting system and design related trade studies outside the control of systems and design engineering. Development program managers sometimes pull systems and design engineers off programs early to save money and then allow procurement, operations or product assurance personnel to conduct trades without the oversight of the appropriate systems or design engineers. This can lead to a multitude of difficulties and usually expensive difficulties. Suffice it to say that systems engineers and design engineers must retain control of systems and design trades throughout the life cycle.

8.2.4 Other Design Trade and Decision Methodologies - The design trade process defined in Figure 8-1 is a proven methodology but not the only useful tool or methodology available. The NASA *Systems Engineering handbook* describes several techniques useful for trade studies and more general decision making. These include:

- **Cost benefit analysis**
- **Influence diagrams/decision trees** – the NASA handbook and Wikipedia have poor descriptions of these tools. A more useful description can be found at http://www.agsm.edu.au/bobm/teaching/SGTM/id.pdf and for a more thorough and mathematical description see http://www.stanford.edu/dept/MSandE/cgi-bin/people/faculty/shachter/pdfs/TeamDA.pdf These tools have available software to facilitate developing the diagrams and converting influence diagrams to decision trees.
- **Multi-criteria decision analysis (MCDA)** - useful for cases where subjective opinions are to be taken into account. Start with: http://en.wikipedia.org/wiki/Multi-criteria_decision_analysis See also http://www.epa.gov/cyano_habs_symposium/monograph/Ch35_AppA.pdf
    - **Analytic hierarchy process (AHP)** – a particular type of MCDA that employs pairwise comparison of alternatives by experts.
- **Utility Analysis** (The DoD SEF calls this Utility Curve Analysis)
    - **Multi-Attribute Utility Theory (MAUT)** (A MCDA technique for Utility Analysis)
- **Risk-Informed Decision Analysis**- for very complex or risky decisions that need to incorporate risk management into the decision process.

8.3 Modeling and Simulation

Modeling and simulation tools are used in all phases of system development from definition to end-of-life. Systems engineers are concerned with the models and simulations used in system definition, design selection and optimization, and performance verification. Systems engineers should identify the models and simulations needed for these tasks during the program planning phase so that any development of required models and simulations can be complete by the time they are needed. Examining the customer's system requirements and the planned trade studies help identify the needed models and simulations. Parameter diagrams are often helpful in identifying the models and simulations needed.

Models constrained by requirements are typically adequate to be used in the system definition phase to define a baseline design concept. Models may be adequate to develop error budgets and allocations but performance simulations are often necessary to select and optimize designs.

System simulations and particularly performance simulations are especially useful in system performance verifications. Therefore it is necessary to include any necessary validation of system simulations in test plans and procedures. End-to-end system simulations are sometimes needed to verify final design compliance with requirements. Other uses include developing the requirements for data analysis tools needed during subsystem and system verification testing, reducing risk and time for developing test software and supporting troubleshooting during test and operational support.

Examples of how models and simulations might be used in system development are shown in Figure 8-4. In this figure the system under development is assumed to be a system that measures parameters by sampling the parameters that are related to a desired phenomenon that cannot be easily or economically measured directly. The measured samples are assumed to be processed first by Data Algorithms, which in this example produce calibrated data. The calibrated data are then input to Product Algorithms which use the calibrated data to produce estimates of the desired phenomena.

It is assumed that a database of truth data is available. This truth data is used in two ways. It is used to predict the parameters that the system is designed to sample by using a model; called a Parameter Model in Figure 8-4. These predicted parameters are then the input to the System Model and System Simulation. The truth data is also used to assess the validity of the system model and the system simulation by comparing the results predicted by the Product Algorithms with the truth data. This example assumes that the System Model generates calibrated data and the System Simulation generates data that must be processed by the Data Algorithms to provide calibrated data. If truth data is available for the desired phenomenon during system operation then the truth data can be used to assess the performance of the system during operation as suggested by the figure.

It is assumed that Environmental Models are developed that can also generate the parameters to be measured. Information from the System Specification is used to generate the parameters in the desired range and with the desired statistics. If the database of truth measurements is representative of the specified range and statistics of the phenomena to be measured then the Parameter Model can be used to generate inputs for system design analysis and as comparisons for system test data analysis and comparisons. If no database of truth measurements is available then Environmental Models are used in place of the Parameter Model but it is not possible to assess results against truth data.

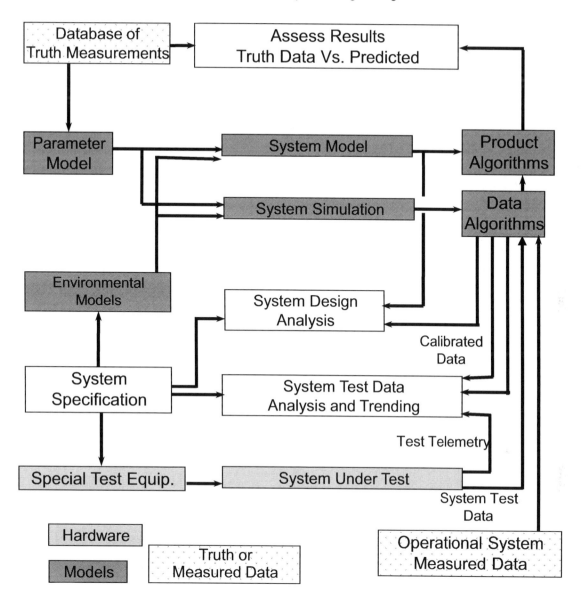

Figure 8-4 Examples of ways models and simulations might be used in developing a sensor or measurement system.

8.3.1 Performance Modeling and Simulation - System performance models and system performance simulations are used in trade studies to evaluate alternative designs and to iteratively optimize the selected design. Typically the system design objective is to develop the **"best value"** design solution. A "best value" design can be defined as:

- Achieves performance above minimum thresholds
- Has life cycle costs within customer's or marketing's defined cost limits
- Meets requirements allocations (mass, power, etc.)
- Assessed to be relatively low risk (so that cost targets are likely attainable)

A general approach to achieving a best value system design is to develop multiple design concepts, assess the cost and performance of each and iterate until the best value is achieved. This usually involves progressively lower level trade studies.

Assessing the cost and performance of system design concepts requires analysis and state-of-the-art tools. Design tools for mechanical, thermal, electrical and optical analysis are well developed, widely available and indispensable for design of modern systems. The same cannot be said for the cost models and top level performance modeling and simulation tools for systems analysis. System performance modeling and simulation tools are too specialized for widespread utility. Thus most systems organizations must develop the modeling and simulation tools needed for defining their systems. Useful **cost models** are available for some systems for organizations developing systems for government agencies like the Defense Department and NASA. Examples of cost estimating models useful for several types of systems and cost estimating tasks include SEER [8-1] and PRICE[8-2].

The first two steps in seeking a best value design are shown in Figure 8-5. The cost model is used to identify a number of design parameters that drive the system cost and quantify how the cost, or the relative cost, depends on each design parameter. The system performance modeling and simulation tools are used to quantify the dependence of system performance on each of the same design parameters.

Having the relationships of cost and performance on design parameters these data can be combined to reveal how the selected design parameters drive the relationship of performance to relative life cycle cost as shown in Figure 8-6. Assuming the cost and performance relationships are determined for n design parameters then the result is n trades of cost vs. performance as a function of each of the n design parameters. It is usually straightforward to select the value of each design parameter that offers the best value design according to the desired criteria. For example, in Figure 8-6 the best value for the design parameter shown is 8 cm because it's near the maximum of the linear portion of the parametric curve and it offers the best performance within the constraints on this particular design parameter.

8.4 Diagrams Useful in Selecting the Preferred Design

As discussed previously much of systems engineering is determining relationships between a system and its environment and among the various subsystems. Ultimately these relationships are defined in detailed drawings but understanding the relationships in order to select the preferred design is aided by examining a system with different levels of abstraction. The modern tools used for electrical and optical design and the design practices of electrical and optical design engineers develop their respective design concepts with diagrams. The diagrams start with block diagrams with a high level of abstraction, perhaps just naming the subsystems, and proceed to greater and greater detail. This process makes it easy for systems engineers and other design engineers to readily understand the electrical and optical design concepts.

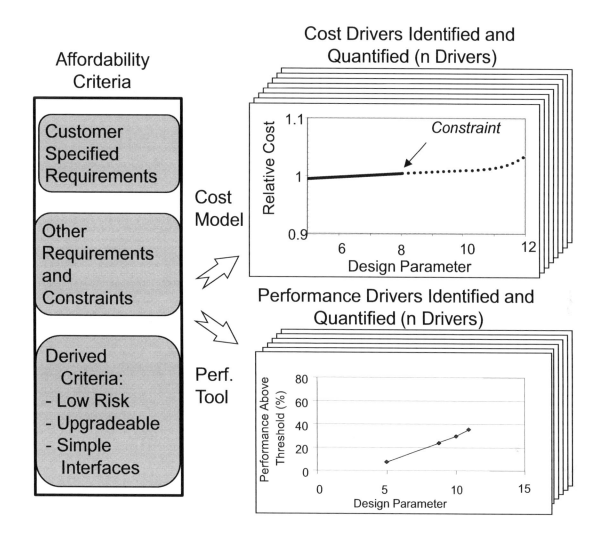

Figure 8-5 Cost models and system performance models and simulations are used to determine the relationship of cost and performance on design parameters.

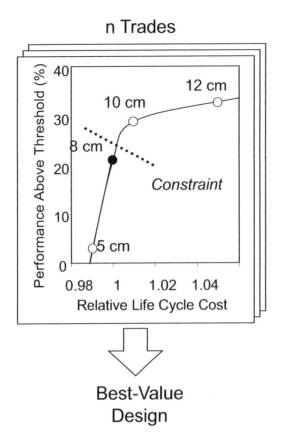

Figure 8-6 The best value design is determined by combining the data from the cost model and performance models and simulations that determine how design parameters drive cost and performance.

Although there are excellent modern design tools providing detailed mechanical and thermal design it's helpful to present the mechanical and thermal design results with high levels of abstraction so that system engineers and other design engineers can easily understand the mechanical and thermal designs. Experienced mechanical and thermal design engineers develop the desired diagrams and tailor their diagrams to the system being developed so that others can readily understand and assess their designs. A few examples are presented here to illustrate how experienced mechanical and thermal designers examine and communicate their design concepts. Note how easy it is to think of alternative design approaches when design concepts are presented in simple block diagram form with a high degree of abstraction. This enables engineers other than expert mechanical and thermal designers to assess design concepts and suggest design alternatives.

8.4.1 Simple Mechanical and Thermal Block Diagrams - Many times a simple block diagram is useful in describing a mechanical or a combined mechanical/thermal design concept. Figure 8-7 is an example showing a system that consists of five assemblies, a frame for mounting the assemblies and a mounting plate that supports the system with a three point mount.

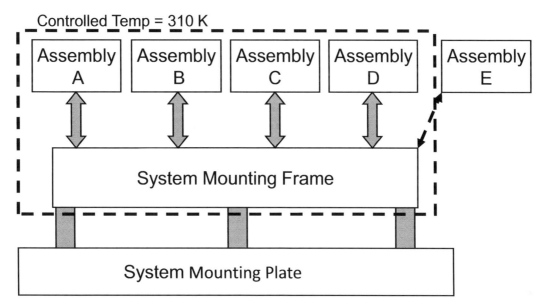

Figure 8-7 A simple mechanical block diagram of a system illustrates how the assemblies interact with each other and the mounting frame.

It's easy to see that the design concept is for each assembly to be coupled to the mounting frame and uncoupled from any other assembly. Four of the assemblies are temperature controlled whereas the fifth assembly is attached to the mounting frame but not within the temperature controlled region. By abstracting all of the size, shape, material and other characteristics of the system the basic mechanical and thermal relationships are easily understood and alternative concepts are obvious. The actual models that the mechanical designer uses to conduct trade studies of alternate mounting concepts are of course much more detailed and usually include the detailed characteristics of the various assemblies, frame and mounting plate; however, this detail is not necessary to explain the concepts and results of the trades to the system engineers and other designers.

Let's suppose that three of the assemblies of the system shown in Figure 8-7 have critical alignment requirements. A perfectly good way to record and communicate the alignment requirements is with allocation trees. However, sometimes it makes the requirements clearer and possibly less prone to misunderstandings if a simple diagram is used in place of a tree. Such a diagram with the alignment requirements for each of the three assemblies and for the attachment points of each on the system mounting frame might look like Figure 8-8.

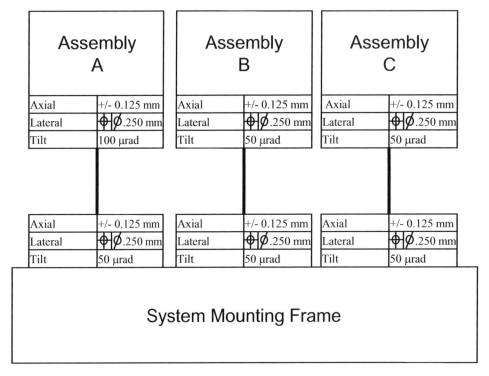

Figure 8-8 A simple block diagram illustrating the alignment requirements for three of assemblies and their respective interfaces with the mounting frame.

Similar simple block diagrams can be used to illustrate temperatures and heat flow paths. It sometimes makes design concepts easier to understand if mechanical and thermal diagrams are developed together. Examples are shown in Figures 8-9 and 8-10 that illustrate simple block diagrams of structural interfaces and thermal interfaces on similar block diagrams.

Even though it takes some time and care to develop diagrams such as shown in these examples the benefits to the team in understanding and refining design concepts to reach the preferred design are well worth the effort. Diagrams like these and related simple diagrams for electrical, optical and other design concepts are invaluable in explaining a design concept to customers and management.

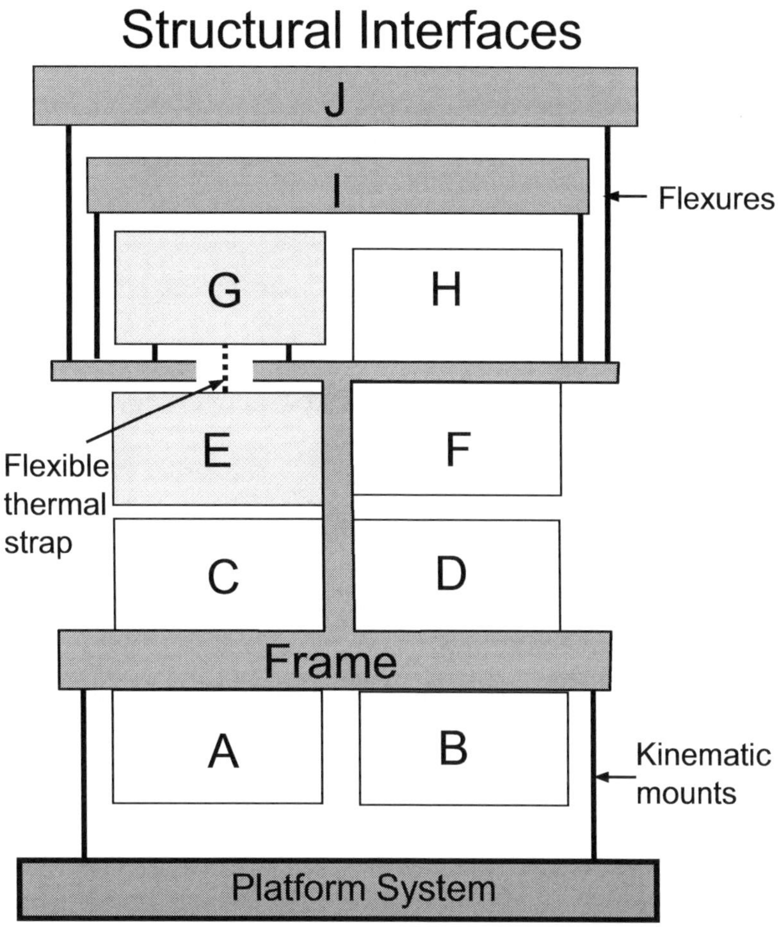

Figure 8-9 A block diagram illustrating structural interfaces within a system and between the system and its parent platform.

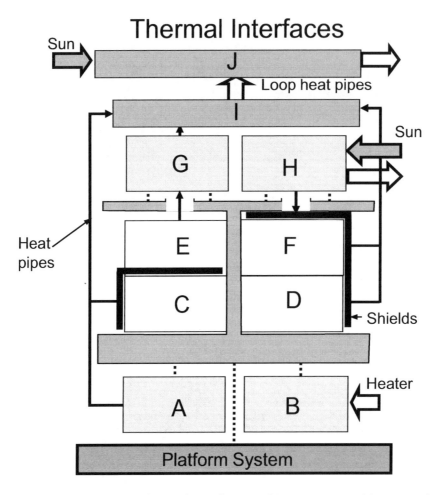

Figure 8-10 A block diagram illustrating thermal interfaces within a system and between the system and its environment using the same diagram approach as used for the structural interface diagram.

8.5 Design Oversight Responsibility

Although it is not the intent of this book to describe systems engineering management processes it is helpful to briefly describe how roles and responsibilities change during the system development phases. The first change in responsibility takes place when the baseline design is refined through trade studies to be the preferred design, e.g. a best value design, and the design requirements database is complete. The system development work is then at the end of the define requirements phase and ready to enter the design phase. This means the design requirements are complete to the level of responsibility of each of the lowest level IPTs, assuming the work is organized so that the lowest level IPT leaders are able to work in the craftsman model, i.e. the leader has the knowledge and experience to make all the design decisions for the work assigned to his/her IPT. At this point the individual IPTs take leadership responsibility from the SEIT or systems engineering and lead in determining how the system is designed and any prototypes built.

During the design phase systems engineering watches over the design to ensure requirements compliance, testability and producibility and monitors MOEs, progress on "ilities" and risk management. In addition, systems engineering is responsible for managing any specification changes. If designers encounter difficulties in meeting an allocated requirement then systems engineers should take responsibility for determining if and how the requirement in question can be modified without jeopardizing overall requirements compliance. The systems engineering role during the design phase can be summarized as supporting the designers to ensure that a balanced and compliant design is achieved that is testable, producible and maintainable.

The transition from the system definition phase to the design phase is typically gated with a design review, e.g. a System Design Review (SDR). Program managers often wish to move systems engineers off a development project when the actions items from a SDR are complete. Although this is a time to increase the design specialty engineers and decrease the systems engineers on the IPTs removing too many systems engineers can leave the designers without adequate systems engineering support and cause other necessary tasks to be understaffed. It is better to assign the systems engineers to tasks like preparing system and subsystem integration and test plans and completing and maintaining the system design documentation.

When the design phase is compete and any prototypes are fabricated then systems engineers resume lead responsibility during test even though design or test engineers may conduct the testing. Typically this change in leadership occurs sometime during integration and subsystem testing of an engineering model or prototype. These leadership responsibility transitions should occur naturally for IPTs with experienced systems and design engineers.

Exercises

I. Review your organizations recent system development programs and determine if efforts were made to improve design concepts via tools like Pugh Concept Selection or if this step was skipped. If your organization is in the practice of going directly to Design Trade Matrices with weighting then introduce the Pugh Concept Selection tool. It is likely to lead to improved design concepts with very little extra effort.

II. Does your organization have a reliable cost modeling tool for the systems it develops? If not review commercially available tools to see if any promise applicability to the systems your organization develops.

III. Assess your systems engineering organization's maturity in modeling and simulation by conducting the following analysis:

    1. Generate a Modeling and Simulation plan for a typical system your organization develops.
        a. Identify the performance models needed
        b. Identify system simulations needed
        c. Develop a flow chart similar to Figure 8-4 showing where each model or simulation is used.
    2. Identify the resources (people, budget and time) needed to develop the models and simulations in your plan.
    3. Can any of the desired models or simulations be adapted from existing capabilities?

4. Can any models or simulations be purchased off the shelf?
5. Are there potential partner or supplier organizations that have existing models or simulations you need or more expertise in developing the models and simulations than your organization?
6. Are your competitors more/less capable in modeling and simulations than your organization?
7. List what actions should be taken by your organization based on your assessment.

9 PROCESSES AND TOOLS FOR VERIFYING TECHNICAL PERFORMANCE

9.0 Introduction

Two important responsibilities of systems engineers are **verification** and **validation**. Verification and validation are formally defined in both the DoD SEF and in the NASA SE handbook. We can simplify these definitions to be verification is answering the question "is the system designed and built right?" and validation is answering the question "is it the right system for the intended purpose?" System engineers must keep these two questions in mind at every step of their work. For example, when a requirement is defined ask "how can the requirement be verified? And ask is this requirement necessary for the system to perform its intended purpose?

There is both a formal and informal aspect to design. The formal aspect typically means analyzing the operation of a system in its intended environment to determine if it performs as intended. This is usually the responsibility of an organization other than the developer, e.g. the customer, or the user for systems developed for the DoD or NASA. Validation of commercial products may be the responsibility of the developer, an independent testing organization or left to the customers. This chapter does not address the formal aspect of system validation. There are other validation activities that are often the responsibility of system engineers, which are defined here but not further discussed. These validation activities include the informal system validation achieved by questioning each requirement as mentioned above and the formal validation of any models, simulations and test equipment and software used in the system development.

In the remainder of this chapter the focus is on verifying the technical performance of the system. There are two approaches to verifying technical performance. One is using good engineering practices in all the systems engineering and design work to ensure that the defined requirements and the design are complete, accurate and meet customer expectations. The other is a formal verification process applied to hardware and software resulting from the design to verify that the system is in compliance with all requirements. Both begin during requirements analysis and continue until a system is operational. The work is represented by the three arrows labeled verification in Figure 6-4 that constitute the backward part of the requirements loop, the design loop and the loop from design synthesis back to requirements analysis and includes both verifying completeness and accuracy of the design and verifying the technical performance of the system.

System verification typically ends with delivery of a system to the customer but can include integration and testing with a higher level system of the customer or another supplier.

9.1 Verifying Design Completeness and Accuracy

Verifying the completeness and accuracy of the design is achieved by a collection of methods and practices rather than a single formal process. The methods and practices used by systems engineers include:

- System engineers checking their own work
- Checking small increments of work via peer reviews
- Conducting formal design reviews
- Using diagrams, graphs, tables and other models in place of text where feasible and augmenting necessary text with graphics to reduce ambiguity
- Using patterns vetted by senior systems engineers to help ensure completeness and accuracy of design documentation
- Developing and comparing the same design data in multiple formats, e.g. diagrams and matrices or diagrams and tables
- Verifying functional architecture by developing a full set of function and mode mapping matrices including mapping:
    - Customer defined functions to functions derived by development team (Some team derived functions are explicit in customer documentation and some are implicit.)
    - Functions to functions for defining internal and external interfaces among functions
    - Sub modes to functions for each of the system modes
- Developing mode and sub mode transition matrices defining allowable transitions between modes and between sub modes
- Using tools such as requirements management tools that facilitate verifying completeness and traceability of each requirement
- Using QFD and Kano diagrams to ensure completeness of requirements and identify relationships among requirements
- Using robust design techniques such as Taguchi Design of Experiments
- Iterating between requirements analysis and functional analysis and between design synthesis and functional analysis
- Employing models and simulations to both define requirements and verify that design approaches satisfy performance requirements
- Validating all models and simulations used in systems design before use (Note the DoD SEF describes verification, validation and accreditation of models and simulations. Here only verification and validation are discussed.)
- Employing sound guidelines in evaluating the maturity of technologies selected for system designs

- Maintaining a through risk management process throughout the development program
- Conducting failure modes and effects analysis (FMEA) and worse case analysis (WCA).

Engineers checking their own work are the first line of defense against human errors that can affect system design performance. One of the most important things that experienced engineers can teach young engineers is the importance of checking all work and the collection of methods for verifying their work that they have learned over their years in the engineering profession. Engineers are almost always working under time pressure and it takes disciple to take the time to check work at each step so that simple human mistakes don't result in having to redo large portions of work. This is the same principle that is behind using peer reviews to catch mistakes early so that little rework is required. Catching mistakes later at major design reviews can result in considerable rework, with significant impact on schedule and budget.

Presenting duplicate methods and tools for the same task in Chapter 6 is not just because different people prefer different methods. It also provides a means of checking the completeness and accuracy of work. The time it takes to develop and document a systems engineering product is a usually a small fraction of the program schedule so that taking time to generate a second version in a different format does not significantly impact schedule and is good insurance against incomplete or inaccurate work.

Pattern based systems engineering, QFD and Taguchi Design of Experiments (DOE) help ensure the completeness, accuracy and robustness of designs. Extensive experience has demonstrated the cost effectiveness of using these methods even though QFD and Taguchi DOE require users to have specialized training to be effective.

Studies[9-1] show that selecting immature technologies result in large increases in costs. Immature technologies often do not demonstrate expected performance in early use and can lead to shortfalls in the technical performance of designs. As a result both NASA and DoD include guidelines for selecting technologies in their acquisition regulations. Definitions of technology readiness levels used by NASA are widely used and are available on the web[9-2]. Definitions are also provided in Supplement 2-A of the DoD SEF.

Failure analysis methods like FMEA and WCA are more often used by design engineers than systems engineers but systems engineers can include failure modes and worse case considerations when defining the criteria used in system design trade studies.

In summary, verifying technical performance during system development includes the disciplined use of good engineering practices as well as the formal performance verification process. This should not be surprising as these practices have evolved through experience specifically to ensure that system designs meet expected performance as well as other requirements.

9.2 Verifying that the System is in Compliance with Requirements

The first phase of verifying that system requirements are met is a formal engineering process that starts with requirements analysis and ends when the system is accepted by its customer. Steps must be taken during the system integration and testing to verify that the system satisfies every "shall" statement in the requirements. These shall statement requirements are collected in a document called the **Verification**

Matrix. The results of the integration and testing of these requirements are documented in a **Compliance Matrix**. The integration and testing is defined and planned in a **System Integration and Test Plan**. Related documentation includes the **Test Architecture Definition**, hardware and software **Test Plans & Procedures** and **Test Data Analysis Plans.**

The roles of systems engineers in verification include:

- Developing the optimum test strategy and methodology and incorporating these into the design as it is developed
- Developing the top level System Integration and Test Plan
- Developing the hierarchy of Integration and Test Plans from component level to system level
- Documenting all key system and subsystem level tests
- Defining system and subsystem level test equipment needed and developing test architecture designs
- Developing the Test Data Analysis Plans
- Analyzing test data
- Ensuring that all shall requirements are verified and documented in the Compliance Matrix.

Good systems engineering practice requires that requirements verification takes place in parallel with requirements definition. The decision to define a requirement with a "shall" or "may" or "should" statement involves deciding if the requirement must be verified and if so how the requirement will be verified. This means that the requirements verification matrix should be developed in parallel with the system requirements documentation and reviewed when the system requirements are reviewed, e.g. at peer reviews and at a formal System Requirements Review (SRR).

9.2.1 Verification Matrix – The verification matrix defines for each requirement the verification method, the level and type of unit for which the verification is to be performed and any special conditions for the verification. Modern requirements management tools facilitate developing the verification matrix. If such tools are not used then the verification matrix can be developed using standard spreadsheet tools.

There are standard verification methods used by systems engineers. These methods are:

1. **Analysis** -Verifies conformance to required performance by the use of analysis based on verified analytical tools, modeling or simulations that predict the performance of the design with calculated data or data from lower level component or subsystem testing. Used when physical hardware and/or software is not available or not cost effective.
2. **Inspection** - Visually verifies form, fit and configuration of the hardware and of software. Often involves measurement tools for measuring dimensions, mass and physical characteristics.
3. **Demonstration** - Verifies the required operability of hardware and software without the aid of test devices. If test devices should be required they are selected so as to not contribute to the results of the demonstration.

4. **Test** - Verifies conformance to required performance, physical characteristics and design construction features by techniques using test equipment or test devices. Intended to be a detailed quantification of performance.

5. **Similarity** - Verifies requirement satisfaction based on certified usage of similar components under identical or harsher operating conditions.

6. **Design** – Used when compliance is obvious from the design, e.g. "The system shall have two modes, standby and operation".

7. **Simulation** – Compliance applies to a finished data product after calibration or processing with system algorithms. May be only way to demonstrate compliance.

The DoD SEF defines only the first four of the methods listed above and these four are sufficient. Many experienced systems engineers find these four too restrictive and also use the other three methods listed. To illustrate a verification matrix with an example consider the function Switch Power. This function might be decomposed as shown in Figure 9-1.

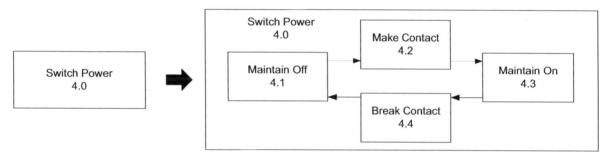

Figure 9-1 A function Switch Power might be decomposed into four sub functions.

An example verification matrix for the functions shown in Figure 9-1 is shown in Figure 9-2. In this example the switch power function is assumed to be implemented in a switch module and that both an engineering model and a manufacturing prototype are constructed and tested. In this example no verification of the switch module itself is specified for production models. Verification of the module performance for production modules is assumed to be included in other system level tests.

It's not important whether the verification matrix is generated automatically from requirements management software or by copy and paste from a requirements spreadsheet. What is important is to not to have to reenter requirements from the requirements document to the verification matrix as this opens the door for simple typing mistakes.

9.2.2 System Integration and Test Plan – The System Integration and Test Plan (SITP) defines the step by step process for combining components into assemblies, assemblies into subsystems and subsystems into the system. It is also necessary to define at what level software is integrated and the levels for conducting verification of software and hardware. Because of the intimate relationship of the verification matrix to the system integration and test the SITP must be developed before the verification matrix can be completed.

	Requirements		Verification				
Para. No.	Item	Requirement	NA	Module	Eng. Mod.	Mfg. Proto.	Comments
4.0	Switch Module	The switch module shall switch a 3 volt supply on and off	x				
4.0.1	Volume	The size shall be 1 x 1 x .25 cm +/- 0.05 cm		I			
4.0.2	Mass	The mass shall be < 1 oz		I			
4.0.3	Power	Power dissapation in the on position shall be < 0.03 watts			T (1)		
4.0.4	Postions	The switch shall have two postions; on and off		Des.			
4.0.5	On/Off Identification	The on and off positions shall be visible to an operator at 100 lumans light level			Dem.		
4.1.6	Contact Pressure	The force to make or break contact without switching shall be > 1.5 lb			T		
4.1.7	Off Resistance	The resistance in the off position shall be > 10E6 Ohms				T	Tests to be done at 0 and 35 C
4.1.8	On Resistance	The resistance in the 0n position shall be < 0.1 Ohms				T	Tests to be done at 0 and 35 C
4.1.9	Switching Force	The force to move the switch from either on or off shall be 1 +/- 0.1 lb			T		
4.1.10	Operability	The switch shall be operable by user wearing cotton gloves				Dem	
(1) If not specified tests are to be conducted at ambient temperature: I = Inspection; Des.= Design; Dem.= Demonstration							

Figure 9-2 An example verification matrix for a switch module.

The SITP defines the buildup of functionality and the best approach is usually to build from the lowest complexity to higher complexity. Thus, the first steps in integration are the lowest levels of functionality; e.g. backplanes, operating systems and electrical interfaces. Then add increasing functionality such as device drivers, functional interfaces, more complex functions and modes. Finally implement system threads such as major processing paths, error detection paths and end-to-end threads. Integration typically happens in two phases: hardware to hardware and software to hardware. This is because software configured item testing often needs operational hardware to be valid. Two general principles to follow are: test functionality and performance at the lowest level possible and, if it can be avoided, do not integrate any hardware or software whose functionality and performance has not been verified. It isn't always possible to follow these principles, e.g. sometimes software must be integrated with hardware before either can be meaningfully tested.

One objective of the SITP is to define a plan that avoids as much as possible having to disassemble the system to implement fixes to problems identified in testing. A good approach is to integrate risk mitigation into the SIPT. For example, there is often a vast difference between the impact of an electrical design problem and a mechanical or optical design problem. Some electrical design or fabrication problems discovered in I & T of an engineering model can be corrected with temporary fixes ("green wires") and I & T can be continued with minimal delay. However, a serious mechanical or optical problem found in the late stages of testing, e.g. in a final system level vibration test, can take months to fix due to the time it takes to redesign and fabricate mechanical or optical parts and conduct the necessary regression testing. Sometimes constructing special test fixtures for early verification of the performance of mechanical, electro-mechanical or optical assemblies is good insurance against discovering design problems in the final stages of I & T.

The **integration plan** can be described with an integration flow chart or with a table listing the integration steps in order. An integration flow chart graphically illustrates the components that make up each assembly, the assemblies that make up each subsystem etc. Preparing the SITP is an activity that benefits from close cooperation among system engineers, software engineers, test engineers and manufacturing engineers. For example, system engineers typically define the top level integration flow for engineering models using guidelines listed above. Manufacturing engineers typically define the detailed integration flow to be used for manufacturing prototypes and production models. If the system engineers use the same type of documentation for defining the flow for the engineering model that manufacturing engineers use then it is likely that the same documentation can be edited and expanded by manufacturing engineers for their purposes.

It should be expected that problems will be identified during system I & T. Therefore processes for reporting and resolving failures should be part of an organizations standard processes and procedures. System I & T schedules should have contingency for resolving problems. Risk mitigation plans should be part of the SITP and be in place for I & T; such as having adequate supplies of spare parts or even spare subsystems for long lead time and high risk items.

System integration is complete when a defined subset of system level functional tests has been informally run and passed, all failure reports are closed out and all system and design baseline databases have been updated. The final products from system integration include the Test Reports, Failure Reports and the following updated documentation:

- Rebaselined System Definition
    - Requirements documents and ICDs
    - Test Architecture Definition
    - Test Plans
    - Test Procedures
- Rebaselined Design Documentation
    - Hardware Design Drawings
    - Fabrication Procedures
    - Formal Release of Software including Build Procedures and a Version Description Document
    - System Description Document
    - TPM Metrics

It is good practice to gate integration closeout with a Test Readiness Review (TRR) to review the hardware/software integration results, ensure the system is ready to enter formal engineering or development model verification testing and that all test procedures are complete and in compliance with test plans. On large systems it is beneficial to hold a TRR for each subsystem or line replaceable unit (LRU) before holding the system level TRR.

9.2.3 Test Architecture Definition and Test Plans and Procedures – The SITP defines the tests that are to be conducted to verify performance at appropriate levels of the system hierarchy. Having defined the tests and test flow it is necessary to define the test equipment and the plans and procedures to be used to conduct the tests. Different organizations may have different names for the documentation defining test equipment and plans. Here the document defining the test fixtures, test equipment and test software is called the **Test Architecture Definition**. The test architecture definition should include the test requirements traceability database and test system and subsystems specifications.

Test Plans define the approach to be taken in each test; i.e. what tests are to be run, the order of the tests, the hardware and software equipment to be used and the data that is to be collected and analyzed. Test Plans should define the entry criteria to start tests, suspension criteria to be used during tests and accept/reject criteria for test results.

Test Procedures are the detailed step by step documentation to be followed in carrying out the tests and documenting the test results defined in the Test Plans. Other terminologies include a **System Test Methodology Plan** that describes how the system is to be tested and a **System Test Plan** that describes what is to be tested. Document terminology is not important; what is important is defining and documenting the verification process rigorously.

Designing, developing and verifying the test equipment and test procedures for a complex system is nearly as complex as designing and developing the system and warrants a thorough systems engineering effort. Neglecting to put sufficient emphasis or resources on these tasks can result in delays of readiness of the test equipment or procedures and risks serious problems in testing due to inadequate test equipment or processes. Sound systems engineering practices treat test equipment and test procedure development as deserving the same disciplined effort and modern methods as used for the system under development.

The complexity of system test equipment and system testing drives the need for disciplined system engineering methods and is the reason for developing test related documentation in the layers of SITP, Test Architecture Definition, Test Plans and finally Test Procedures. The lower complexity top level layers are reviewed and validated before developing the more complex lower levels. This approach abstracts detail in the top levels making it feasible to conduct reviews and validate accuracy of work without getting lost in the details of the final documentation.

The principle of avoiding having to redo anything that has been done before also applies to developing the **Test Architecture Definition, Test Plans** and **Test Procedures**. This means designing the system to be able to be tested using existing test facilities and equipment where this does not compromise meeting system specifications. When existing equipment is inadequate then strive to find commercial off the shelf (COTS) hardware and software for the test equipment. If it is necessary to design new special purpose test equipment then consider whether future system tests are likely to require similar new special purpose designs. If so it may be wise to use pattern base systems engineering for the test equipment as well as the system.

Where possible use test methodologies and test procedures that have been validated through prior use. If changes are necessary developing Test Plans and Procedures by editing documentation from previous system test programs is likely to be faster, less costly and less prone to errors than writing new plans. Sometimes test standards are available from government agencies.

9.2.4 Test Data Analysis – Data collected during systems tests often requires considerable analysis in order to determine if performance is compliant with requirements. The quantity and types of data analysis needed should be identified in the test plans and the actions needed to accomplish this analysis are to be included in the test procedures. Often special software is needed to analyze test data. This software must be developed in parallel with other system software since it must be integrated with test equipment and validated by the time the system completes integration. Also some special test and data analysis software may be needed in subsystem tests during integration. Careful planning and scheduling is necessary to avoid project delays due to data analysis procedures and software not being complete and validated by the time it is needed for system tests.

9.2.5 Compliance Matrix – The data resulting from the actions summarized in the verification matrix for verifying that the system meets all requirements are collected in a **compliance matrix**. The compliance matrix shows performance for each requirement. It flows performance from the lowest levels of the system hierarchy up to top levels. It identifies the source of the performance data and shows if the design is meeting all requirements. The bottom up flow of performance provides early indication of non-compliant system performance and facilitates defining mitigation plans if problems are identified during verification actions. An example compliance matrix for the switch module is shown in Figure 9-3.

Note that the requirements half of the compliance matrix is identical to the requirements half of the verification matrix. The compliance matrix is easily generated by adding new columns to the verification matrix. Results that are non-compliant, such as the switching force, or marginally compliant, such as the on resistance, can be flagged by adding color to one of the value, margin or compliant columns or with notes in the comments column.

In summary, the arrows labeled verification in Figure 6-4 from functional analysis to requirements analysis, from design to functional analysis and from design to requirements analysis relate to the iteration that the systems engineers do to ensure the design is complete and accurate and that all "shall" requirements are verified in system integration and system test. This iteration is necessary so that for each requirement a verification method is identified, any necessary test equipment, test software and data analysis software is defined in time to have validated test equipment, test procedures and test data analysis software ready when needed for system integration and test.

Requirements			Compliance			
Para. No.	Item	Requirement	Value	Margin	Compliant	Comments
4.0	Switch Module	The switch module shall switch a 3 volt supply on and off				
4.0.1	Volume	The size shall be 1 x 1 x .25 cm +/- 0.05 cm	0.95 x 0.98 x 0.27	(-.05 x -.02 x +.02)	Yes	
4.0.2	Mass	The mass shall be < 1 oz	0.96	0.04	Yes	
4.0.3	Power	Power dissapation in the on position shall be < 0.03 watts	0.029	0.001	Yes	Test conducted at 20 C
4.0.4	Postions	The switch shall have two postions; on and off	Yes		Yes	
4.0.5	On/Off Identification	The on and off positions shall be visible to an operator at 100 lumans light level		Yes	Yes	Demonstrated at 90 lumens
4.1.6	Contact Pressure	The force to make or break contact without switching shall be > 1.5 lb	1.9	0.4	Yes	Per test procedure TP 23.4
4.1.7	Off Resistance	The resistance in the off position shall be > 10E6 Ohms	1.65 x 10e6 at 35 C	5 e5 at 0 C 65 e5 at 35 C	Yes	
4.1.8	On Resistance	The resistance in the On position shall be < 0.1 Ohms	0.098 at 0 C 1.000 at 35 C	.002 at 0 C; 0 at 35 C	Yes	
4.1.9	Switching Force	The force to move the switch from either on or off shall be 1 +/- 0.1 lb	1.13	0.03	No	
4.1.10	Operability	The switch shall be operable by user wearing cotton gloves	Yes		Yes	Female Operator wearing jersey gloves

Figure 9-3 An example Compliance Matrix for the simple switch function illustrated in Figure 9-1.

9.3 Systems Engineering Support to Integration, Test and Production

Manufacturing personnel and test personnel may have primary responsibility for integration, test and production however; systems engineers must provide support to these tasks. Problem resolution typically involves both design and systems engineers and perhaps other specialty engineers depending on the problem to be solved. Systems engineers are needed whenever circumstances require changes in parts or processes to ensure system performance isn't compromised.

Exercises:

1. Why can't developing the requirements verification matrix and developing test equipment be done during detailed design?
2. Should developing the system test equipment be assigned to junior engineers? Why?
3. Should every system level requirement be a "shall" statement?
4. Why shouldn't systems engineers be removed from development programs as soon as all requirements documentation is complete?
5. Why is it beneficial in some cases to include similarity, design and simulation as verification methods?

10 PROCESSES AND TOOLS FOR RISK AND OPPORTUNITY MANAGEMENT

10.0 Introduction

Risk is always present; its presence is a fact of nature. Accepting that risk is always present is the first step toward managing risks to reduce their effects. Managing risk is the responsibility of the development program leaders but the mechanics are often delegated to systems engineering. Even if systems engineers are not responsible for maintaining the processes and tools it is essential that they understand the importance of risk management and the methods used for effective risk management. Inattention to risk management is the second highest cause of projects not meeting expectations. Effective risk management is an iterative process to be performed throughout the product development effort. Just like other systems engineering processes it takes experience and discipline to conduct effective risk management.

Development programs also have **opportunities** for improving cost, schedule or system performance. It is important to identify and manage opportunities as well as risks in order to have an effective program. This chapter defines risk, outlines a risk management process that can be used for risk and opportunity management and provides examples of templates and processes useful for risk and opportunity management.

10.1 Risk Definition

Risk is the consequence of things happening that negatively impact the performance of a system development project. Risks arise from events that occur inside and outside the development organization. The consequence of an event can impact the quality, cost or schedule of a system development project, or some combination of these effects. There are risks in any project but there are usually more risks associated with projects that are new to the development organization's experience. Risks are always present in the development of new products or services or changes to the processes, people, materials or equipment used in the development of products or services. Risks to developing new products and services arise from unplanned changes to the internal environment or changes in the external environment, such as the economy, costs of materials, labor market, customer preferences or actions by a competitor, a regulating body or a government agency. An effective development team faces up to risks and manages risks so that the negative impacts are minimized.

There is an operational definition of risk that aids in managing risk. This definition is:

Risk R is The Probability p of an Undesirable Event Occurring; Multiplied by The Consequence of the Event Occurrence measured in arbitrary units C or dollars \$; R=p x C or R=p x \$.

This definition allows risks to be quantified and ranked in relative importance so that the development team knows which risks to address first, i.e. the risks with the highest values of R. If the event consequence is measured in dollars then it's easier to evaluate how much budget is reasonable to assign to eliminate or reduce the consequence of the risk.

The second definition measures risk in units of dollars. Thus impacts to the quality of a product or service or to the schedule of delivering the product or service are converted to costs. Impacts to quality are converted to dollar costs via estimated warranty costs, cost of the anticipated loss of customers or loss of revenue due to anticipated levels of discounting prices. Schedule delays are converted to dollar costs by estimating the extra costs of labor during the delays and/or the loss of revenue due to lost sales caused by the schedule delays.

Opportunities can also be defined operationally by the product of the probability an opportunity for improvement can be realized and the consequence if the opportunity is realized, measured either in arbitrary units or dollars. In the rest of this chapter when risk is addressed the reader should remember that opportunities can be treated by the same methods as risks.

The key to good risk management is to address the highest risk first. There are three reasons to address the highest risk first. First is that mitigating a high risk can result in changes to plans, designs, approaches or other major elements in a project. The earlier these changes are implemented the lower the cost of the overall project because money and people resources are not wasted on work that has to be redone later. The second reason is that some projects may fail due to the impossibility of mitigating an inherent risk. The earlier this is determined the fewer resources are spent on the failed project thus preserving resources for other activities. The third reason is that any project is continually competing for resources with other activities. A project that has mitigated its biggest risks has a better chance of competing for continued resource allocation than activities that still have high risks.

10.2 Managing Risk

Managing risk means carrying out a systematic process for identifying, measuring and mitigating risks. Managing risk is accomplished by taking actions before risks occur rather than reacting to occurrences of undesirable events. The DoD SEF defines four parts to risk management and the NASA SE Handbook defines five top level parts and a seven block flow chart for risk management. It is helpful to decompose these into 11 steps. The 11 steps in effective risk management are:

1. Listing the most important requirements that the project must meet to satisfy its customer(s). These are called Cardinal Requirements and are identified in requirements analysis or via Quality Function Deployment

2. Identifying every risk to a project that might occur that would have significant consequence to meeting each of the Cardinal Requirements

3. Estimating the probability of occurrence of each risk and its consequences in terms of arbitrary units or dollars

4. Ranking the risks by the magnitude of the product of the probability and consequence (i.e. by the definition of risk given above)

5. Identifying proactive actions that can lower the probability of occurrence and/or the cost of occurrence of the top five or ten risks

6. Selecting among the identified actions for those that are cost effective

7. Assigning resources (funds and people) to the selected actions and integrating the mitigation plans into the project budget and schedule

8. Managing the selected action until its associated risk is mitigated

9. Identifying any new risks resulting from mitigation activities

10. Replace mitigated risks with lower ranking or new risks as each is mitigated

11. Conduct regular (weekly or biweekly) risk management reviews to:

    - Status risk mitigation actions
    - Brainstorm for new risks
    - Review that mitigated risks stay mitigated.

In identifying risks it is important to involve as many people that are related to the activity as possible. This means people from senior management, the development organization, other participating organizations and supporting organizations. Senior managers see risks that engineers do not and engineers see risks that managers don't recognize. It is helpful to use a list of potential sources of risk in order to guide people's thinking to be comprehensive. A list might look like that shown in Figure 10-1.

It also helps ensure completeness of understanding risks if each risk is classified as a technical, cost or schedule risk or a combination of these categories.

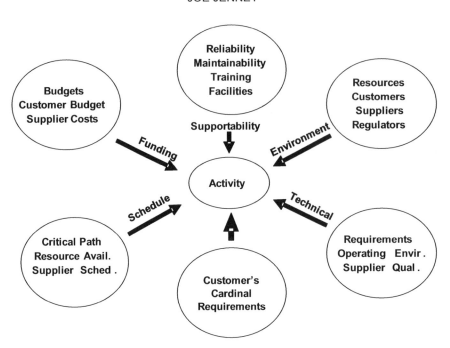

Figure 10-1 An example template for helping identify possible sources of risk to the customer's cardinal requirements.

10.3 Tools for Risk Management

Standard tools for risk management include risk matrices; also called risk summary grids, and risk registers. There are also tables of definitions and guidelines that aid in using the matrices and registers. A methodology useful for reducing risk through proactive and planned build and test steps is called design iteration. These tools and design iteration are described in this chapter. Other tools aiding or supporting the identification of risks include fault trees, worst case analysis and failure modes analysis. Risk burn down charts that display how the total expected value of all identified risks is reduced with time as mitigation actions are completed are useful in monitoring the overall progress of risk mitigation and the effectiveness of budgeting for risk management.[10-1]

10.3.1 Risk Summary Grid - The risk summary grid is a listing of the top ranked risks on a grid of probability vs. impact. The risk summary gird is excellent for showing all top risks on a single graphic and grouping the risks as low, medium or high. Typical grids are 3 x 3 or 5 x 5. An example 5 x 5 template is shown in Figure 10-2.

The 5 x 5 risk summary grid enables risks to be classified as low, medium or high; typically color coded green, yellow and red respectively, and ranked in order of importance. Note that the definitions for low and medium are not standard. The definition used in Figure 10-2 is conservative in limiting low risk to the five squares in the lower left of the grid with risk values of 0.5 or less. Medium risks have values of 0.7 to 3.5 and high risks have values from 4.5 to 8.1. Others, e.g. the *Risk Management Guide for DOD Acquisi-*

tion[10-2], an excellent tutorial on risk management, define the entire first column plus six other lower left squares as low risk.

Probability	0.9	0.9	2.7	4.5	6.3	8.1
	0.7	0.7	2.1	3.5	4.9	6.3
	0.5	0.5	1.5	2.5	X 3.5	4.5
	0.3	0.3	0.9	1.5	2.1	2.7
	0.1	0.1	0.3	0.5	0.7	0.9
		1	3	5	7	9
		Impact				

Figure 10-2 An example of a 5 x 5 risk summary grid template

Relative importance is the product of probability and impact. Identified risks are assigned to a square according to the estimates of their probability of occurrence and impact to the overall activity. In Figure 10-2 there is one medium risk, shown by the x in the square with a probability 0.5, impact 7 and therefore having a relative importance of 3.5. The numbers shown for impact are arbitrary and must be defined appropriate to the activity for which risk is being managed.

Some risk management processes described on the web use letters rather than numbers to rank risk probability in constructing risk summary grids. The objective is to assign either a probability number or letter to each risk. To do this it is necessary to make a judgment of the likelihood that the risk occurs. The table shown in Figure 10-3 provides reasonable guidelines for such judgments. Thus, if the likelihood of an event occurring is judged to be remote then assign the probability of 0.1 or the letter A. If it is highly likely assign 0.7 or D. It can be argued that guidelines are needed for what is remote or likely. Unfortunately this wouldn't help as there is always some guess work or judgment required. If several members of a team discuss the likelihood then they can probably reach agreement and this is adequate. It is important for the novice to understand that it isn't essential that the probabilities are exact. The objective is to come close enough to compare the relative probabilities of several events so that the events can be prioritized in relation to their relative risk or relative probability of occurrence.

Criteria	Probability	Probability
Remote	0.1	A
Unlikely	0.3	B
Likely	0.5	C
Highly Likely	0.7	D
Near Certainty	0.9	E

Figure 10-3 Guidelines for assigning probability numbers or letters to risk based on judgment criteria.

After assigning a probability to a risk it is necessary to make a judgment of the impact of occurrence of the risk. A risk event can cause an unexpected cost or cost increase, a slip in the schedule for achieving some related event or reduce the quality or technical performance of some design requirement. It is also possible for the risk to impact two or even all three of the cost, schedule or quality measures. The table shown in Figure 10-4 provides one set of guidelines for assigning impact numbers 1, 2, 3, 4 or 5 to a risk event.

Technical	Schedule	Cost	Impact
N.A.	N.A.	N.A.	0
Minimal	Minimal	Minimal	1
Acceptable with some reduction in margin	Can meet with additional resources	< 5%	2
Acceptable with large reduction in margin	Minor slip in key milestones	5-7%	3
Acceptable, no margin	Major slip in key milestones or Critical Path impacted	7-10%	4
Unacceptable	Can't achieve key milestone	>10%	5

Figure 10-4 Guidelines for assigning impact numbers to a risk event.

A risk summary grid template using the guidelines provided in Figures 10-3 and 10-4 is shown in Figure 10-5.

Probability	Risk Level				
E=90%	Medium-0.9	Medium-1.8	High-2.7	High-3.6	High-4.5
D=70%	Low-0.7	Medium-1.4	Medium-2.1	High-2.8	High-3.5
C=50%	Low-0.5	Medium-1.0	Medium-1.5	Medium-2.0	High-2.5
B=30%	Low-0.3	Low-0.6	Medium-0.9	Medium-1.2	Medium-1.5
A=10%	Low-0.1	Low-0.2	Low-0.3	Low-0.4	Low-0.5
	1	2	3	4	5

Figure 10-5 A less conservative risk summary grid template using the guidelines provided in Figures 10-3 and 10-4.

The process using a 3 x 3 risk summary grid typically assigns probability of risks as 0.1, 0.3 or 0.9 and impacts as 1, 3 or 9. There are three squares for each of the low, medium and high risk classifications with relative importance values ranging from 0.1 to 8.1 according to the products of probability and impact. An example of a 3 x 3 risk summary grid template is shown in Figure 10-6.

Probability				
	0.9	Med. 0.9	High 2.7	High 8.1
	0.3	Low 0.3	Med. 0.9	High 2.7
	0.1	Low 0.1	Low 0.3	Med. 0.9
		1	3	9
		Impact		

Figure 10-6 An example template for a 3 x 3 risk summary grid.

Specific process details or numerical values are not important. What is important is having a process that allows workers and managers to assess and rank risks and to communicate these risks to each other, and in some cases to customers. The simple risk summary grids are useful tools for accomplishing these objectives and are most useful in the early stages of the life cycle of an activity and for communicating an overall picture of risks.

The identified risks are collected in a list and the ten or so with the highest risk values are numbered or given letter identifications. The associated numbers or letters are then displayed in the appropriate square on the risk summary grid. In use the risk values of each square are either not shown in the square or made small so there is room for several risk identifiers in a square. The risk summary grid then provides a quick visual measure of the number of high, medium and low risks. In the early stages of a project it should be expected that there are more risks in the high and medium categories than the low and as risk mitigation progresses the number of high risks are reduced.

Having identified the risks and ranked them the team must decide what to do with risks that are assigned as Low, Medium or High. One set of guidelines is shown in the table provided in Figure 10-7.

Risk Value	Risk Level	Definition
2.5 - 4.5	High	UNACCEPTABLE- Major disruption likely. Change approach. Management decision required
0.9 - 2.1	Medium	MODERATE- Some disruption. Consider alternate approach. Management attention necessary.
0.1 - 0.7	Low	LOW- Minimal impact. Oversight required to ensure risk remains low.

Figure 10-7 Example guidelines for actions for each level of risk.

Again, the specific guidelines a team employs is not as important as it is for the team to have agreed upon guidelines appropriate to their work and organization and to follow them.

10.3.2 Risk Register - The risk summary grid can be used as a tool in the development team's ongoing risk management meetings but a better tool is the risk register. The risk register ranks risks by the expected dollar value of each risk according to the operational definition of risk given earlier. Constructing the risk register on a spreadsheet allows risks to be sorted by dollar value so that the highest risks are always on top of the list. The risk register also facilitates keeping all risks in the same data base even though management actions may be active on only the top five or ten at any time. When a high risk is mitigated the expected dollar value of the risk is reduced and it typically falls out of the top five or ten but is still on the list. This enables reviewing mitigated risks to ensure they remain mitigated or to readdress a risk at a later time when all the higher risks have been mitigated to even lower values. An example of a simple risk register with three risks constructed on a spread sheet is shown in Figure 10-8.

The risk type and impact if risk occurs are usually described as "if", "then" statements as in Figure 10-8. This helps the management team remember specifically what each risk entails as they conduct reviews over the life of the activity. Expected values are expressed in dollars, which facilitates both ranking and decisions about how much resources should be assigned to mitigation activities. Assuming of course that in managing activities in the development organization it is the practice to hold some fraction of the budget in reserve to handle unforeseen events. Funds from this reserve budget are assigned to risk mitigation activities. Risk mitigation actions should be budgeted and scheduled as part on on-going work. A failure many inexperienced managers make is handling risks outside of the mainline budget and schedule. This undisciplined approach often leads to risk management degenerating into an action item list and finally to a reactive approach to unexpected events rather that a proactive approach to reduce the risks systematically.

Risk	Impact if Risk Occurs	Cost of Occurrence	Probability of Ocurrence	Expected Cost ($)	Mitigation Plan	Responsible Person
If Processor chip delivery late	Then day for day schedule slip	Six Week Slip = $50,000	0.7	$35,000	Incentivise early delivery	John Doe
If data rate exceeds 2 Mbps	Then require interface module redesign	$60,000	0.5	$30,000	Investigate lossless data compression	Mary Smith
If interrupt module power exceeds 1 w	Then larger battery required	#100 per unit = $90000	0.3	$27,000	Breadboard two alternative ckt's	Sam Jones

Figure 10-8 An example of a risk register constructed in columns on a spread sheet.

A more complete risk register template than the example shown in Figure 10-8 might contain columns for the risk number, title, description (if), impact (then), types (three columns: cost, schedule, quality or technical), probability of occurrence, cost impact, schedule impact, mitigation plan and mitigation schedule. The form of the risk register template is not critical so the team managing the risks should construct a template that contains the information they feel they need to effectively manage risks.

The risk register, if properly maintained and managed, is a sufficient tool for risk management on small and short duration projects. Setting aside an arbitrary management reserve budget to manage risks is ok for small projects. Portions of the reserve are allocated to mitigation of risks and the budgets and expenses for risk mitigation can be folded into the overall cost management system. Large, long duration projects or high value projects warrant a more focused approach to budgeting for risk management. For example, one approach is to track the planned and actual mitigation expense and the dollar value of remaining risks as a function of time. These management actions do not usually involve systems engineering but systems engineers should be aware of the methods used for management of risk reduction budgets.

In summary, spending a small amount of money in proactively mitigating risks is far better than waiting until the undesirable event occurs and then having to spend a large amount of money fixing the consequences. Remember that risk management is <u>proactive</u> (problem prevention) and not reactive. Also risk management is <u>NOT</u> an action item list for current problems. Finally, risk management is an on-going activity. Do not prepare risk summary grids or risk registers and then put them in a file as though that completes the risk management process, a mistake inexperienced managers make too often.

10.4 Design Iterations Reduce Risk

Design iterations are planned "build, test and learn" activities for high risk parts of the system; parts that are small enough to build, test and assess rapidly. Examples best illustrate the concept of using design iterations to reduce risk:

- Engineering builds and tests two types of breadboard circuits to get data needed for a subsystem specification, trade studies and subsequent detailed design

- Manufacturing pilots a new production process during architecture definition and uncovers yield problem early

- Analysts simulate three candidate signal processing algorithms during concept development and recommend the best

- Software implements and tests high risk parts of three alternative approaches for system control software during requirements analysis.

Design iteration is not a fire fighting technique; it is a methodical risk reduction methodology. Design iteration is not "build the entire system, test it and fix it if it doesn't work" approach; in fact it is intended to avoid falling into such an unproductive approach. Note that "spiral development", described earlier, is system level risk reduction. Design iterations can be thought of as a methodology that supports implementing progressive freeze.

Recall the message in Figure 6-32 that the cost of making design changes is low in the early stages of a system development when there are many degrees of freedom and becomes higher as the development progresses and there are fewer and fewer degrees of freedom. Thus design iterations are cost effective for many high risk items in the early stages of a development. It's ok to have many short cycle iterations in parallel in early phases and it's ok to "throw away" some results as the team learns and lowers risk.

Exercise

1. Does your organization have a standard risk management process in place? If so then is it being effectively used on current development programs? If not then think through a plan to put a standard process in place and train workers to use it. This can be a commercial process or a process you or your team develops. Implement it via formal training or on an incremental basis. The important thing is having a process and using it religiously.

11 INTRODUCTION TO MODEL BASED SYSTEMS ENGINEERING

11.0 Introduction

The advantages of using labeled graphical models, diagrams, tables of data and similar non prose descriptions compared to natural language or prose descriptions have been discussed several times. Now we make a distinction between two types of models. One type is, as stated, a non-prose description of something. The second type is analysis models; either static models that predict performance or dynamic models referred to as simulations. Static analysis models may be strictly analytical or may be machine readable and executable. Modern simulations are typically machine readable and executable. This is an arbitrary distinction as the DoD defines a model as a physical, mathematical, or otherwise logical representation of a system, entity, phenomenon, or process. (DoD 5000.59 -M 1998)

In Chapter 5 it was stated that PBSE is model based but includes prose documents as well. The models used in PBSE can be either the first type or the second type. Now we want to introduce a different approach to using models for systems engineering. This approach is called **Model Based System Engineering** (MBSE) and it strives to accomplish system engineering with models that are machine readable, executable or operative. An INCOSE paper[11-1] defines MBSE as an approach to engineering that uses models as an integral part of the technical baseline that includes the requirements, analysis, design, implementation, and verification of a capability, system, and/or product throughout the acquisition life cycle.

This chapter is an introduction to MBSE; no attempt is made to review or even summarize the extensive literature on MBSE. MBSE is rapidly evolving, facilitated both by development of commercial tools and by an INCOSE effort to extend the maturity and capability of MBSE over the decade from 2010 to 2020. Whereas we attempt to describe how MBSE offers benefits compared to traditional prose based systems engineering it isn't claimed that pure MBSE is superior or inferior to methodologies that mix MBSE, PBSE and traditional methods. The intent is to provide an introduction that enables readers to assess how MBSE can be beneficial to their work and to point the way toward further study.

Traditional systems engineering is a mix of prose based material, typically requirements and plans, and models such as functional diagrams, physical diagrams and mode diagrams. Eventually design documentation ends in drawings, which are models. MBSE can be thought of as replacing the prose documents that define or describe a system, such as requirements documents, with models. We are not concerned as much with plans although plans like test plans are greatly improved by including many diagrams, photos and other models with a minimum of prose.

To some it may seem difficult to replace requirements documents with models. However, QFD can be stand-alone systems engineering process and QFD is a type of MBSE. Although it does not attempt to heavily employ machine readable and executable models, QFD is an example of defining requirements in the form of models. Another way to think about requirements is that mathematically requirements are graphs and can therefore be represented by models. A third way to think about requirements as models is as tree structures. Each requirement may have parent requirements and daughter requirements and just as no leaf of a tree can exist without connection to twigs, twigs to limbs, and limbs to the trunk no requirement can stand alone. Trees can be represented by diagrams so requirements can all be represented in a diagram.

Throughout this book there is an emphasis on representing design information as models in order to reduce ambiguity and the likelihood of misinterpretation of text based design information. There is also an emphasis on using analysis models and simulations as much as possible throughout the life cycle of a system development. The use of models and simulations improves functional analysis, design quality, system testing and system maintenance. Think of MBSE as combining these two principles; then it becomes clear why MBSE is desirable. Another way to look at traditional systems engineering vs. MBSE is for traditional systems engineering engineers write documents and then models are developed from the documents. In MBSE the approach is to model what is to be built from the beginning.

Model based design has been standard practice for many engineering specialties since the 1980s. Structural analysis, thermal analysis, electrical circuit analysis, optical design analysis and aerodynamics are a few examples of the use of Computer Aided Design (CAD) or model based design analysis. It is systems engineering that been slow to transition from non-model based methods, with the exception of performance modeling and simulation. To achieve the benefits of MBSE systems engineers need to embrace requirements diagrams, Use Case analysis and other MBSE tools along with performance modeling and simulation.

11.1 Definitions of Models As Applied to MBSE

Models have been referred to throughout this material without providing a formal definition or defining the types of models typically used in systems engineering. Formally, a model is a representation of something, as described in the DoD definition given above. For our purposes a model is a representation of a design element of a system. Types of models of interest to MBSE include[11-2]:

Schematic Models: A chart or diagram showing relationships, structure or time sequencing of objects. For MBSE schematic models should have a machine-readable representation. Examples include FFBDs, interface diagrams and network diagrams.

Performance Model: An executable representation that provides outputs of design elements in response to inputs. If the outputs are dynamic then the model is called a simulation.

Design Model: A machine interpretable version of the detailed design of a design element. Design models are usually represented by CAD drawings, VHDL, C, etc.

Physical model: A physical representation that is used to experimentally provide outputs in response to inputs. A breadboard or brass board circuit is an example.

Achieving machine readable and executable models means that the models must be developed using software. Useful languages used by software and systems engineers for such models are the Unified Modeling Language™ (UML®) and its derivative SysML™. A brief introduction to these languages is presented here along with references for further study.

11.2 Introduction to the Unified Modeling Language™(UML®)[1]

The Unified Modeling Language is a language used to specify, visualize, and document models of software systems, including their structure and design. The language was formed by Grady Booch, Ivar Jacobson, and James Rumbaugh[11-3] after a critical mass of ideas started forming in the 1990s from their individual work in object-oriented methods. Collectively they defined UML for three reasons:

1. Their methods were evolving toward each other independently
2. To provide stability to the object-oriented modeling language
3. To provide improvements to the language

UML 1.0 was offered by the Object Management Group (OMG) for standardization in January 1997 with the final acceptance of version 1.1 in September 1997. The OMG maintains the UML specification with the latest version found on the OMG website at **http://www.uml.org/**

In March 2003, the OMG along with the International Council on Systems Engineering (INCOSE) developed a Request for Proposal (RFP) for UML for Systems Engineering. The RFP specified the requirements for extending UML to support the needs of the systems engineering community. **SysML™**[1] is a general purpose graphical modeling language for specifying, analyzing, designing and verifying complex systems. The system may include hardware, software, information, personnel, procedures, and facilities. SysML represents a subset of UML. The SysML specification was developed with the OMG announcing the adoption of the OMG SysML in July 2006 and the availability of the specification in September 2007. The latest version of the SysML specification can be found at the OMG website at **http://www.omg.org/spec/SysML/**

11.2.1 Types of Diagrams in the UML- UML 2.4 (**http://www.omg.org/spec/UML/2.4**) defines fourteen types of diagrams, divided into two categories:

1. Structure Diagrams, which include:

---

[1] Systems Modeling Language and SysML are either registered trademarks or trademarks of Object Management Group, Inc. in the United States and/or other countries.

a. Class Diagram
   b. Package Diagram (diagram same for SysML)
   c. Object Diagram
   d. Component Diagram
   e. Composite Structure Diagram
   f. Deployment Diagram
2. Behavior Diagrams, which include:
   a. Activity Diagram
   b. Sequence Diagram (diagram same for SysML)
   c. Communication Diagram
   d. Timing
   e. Timing Diagram (optional)
   f. Interaction Tables (optional)
   g. State Machine Diagram (diagram same for SysML)
   h. Use Case Diagram (diagram same for SysML)

Structure diagrams are used to visualize, specify, construct and document the static aspects of a system (i.e. no time element). **Behavior diagrams** are used to visualize, specify, construct and document the dynamic aspects of a system (i.e. time is considered).

Not all diagrams in each category need to be used when defining a software system, just what is needed to communicate how the software system works; how it is built structurally, how it behaves, and what it interacts with. Without both category representations, a software system is not fully defined and can lead to issues when the software system is built. For example, without understanding the timing requirements for a real-time embedded system, software designers may define the software structure to meet all functional requirements of the system, but find out during testing that it does not meet the timing requirements. The software system would need to be re-architected to meet both the functional and timing requirements.

The next sections provide simple examples of some of the UML diagram types including common diagrams with SysML. These are simple examples based on the Digital Camera System first introduced in Chapter 6. Note that the examples are not meant to represent a complete Digital Camera System or to be 100% accurate. They are just representative of the UML diagrams to help understand the basic structure and information they are conveying. There are many books that provide more detailed instructions on how to create these diagrams and on what they are used for when modeling a software system. There are also many Web Sites that provide examples of each diagram type.

11.2.2 Structure Diagrams- Examples of Class Diagrams, Package Diagrams and Deployment Diagrams are Structure Diagrams shown and discussed in this section.

Class Diagram - A class diagram is used to show a set of classes, interfaces, and collaborations and their relationships. Class diagrams provide a static view of the system. Using the Simple Digital Camera System, Figure 11-1 illustrates a possible class diagram based on Take a Digital Picture Use Case.

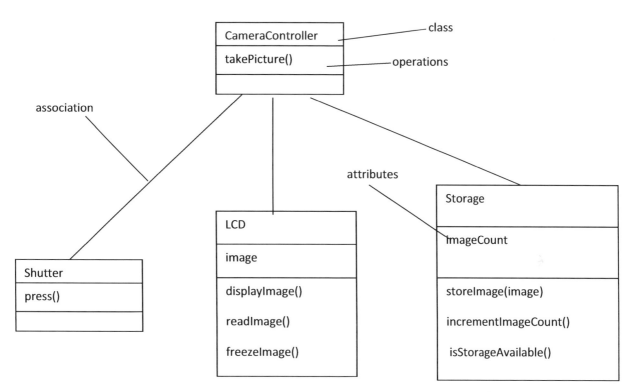

Figure 11-1 Class Diagram Example for a Simple Digital Camera System

Shutter, LCD, and Storage are all classes within the system that interact with each other and represent real world items that make-up a Digital Camera. Each class contains **attributes** and **operations**. Attributes are a named property of a class that describes a range of values that instances of the property may hold; using software terms, a data structure or a variable. An attribute is some piece of information the class needs in order to perform its function. A class may have any number of attributes or none at all. The class Storage has an attribute of imageCount to keep track of the number of images stored and it has three operations indicated, storeImage, incrementImageCount, and isStorageAvailable.

When first identifying classes within a system start with the requirements specification or Use Cases and identify nouns. Alternatively, start with the real world items that make-up the system. All of these are candidates for possible classes. As more is understood about the system, the class diagrams will evolve until the final class diagram for the system is achieved. In modeling a software system, the classes tend to become more abstract and may not have real world counterparts.

The example above does not contain all possible graphic elements that are possible in a class diagram. For more information, see the UML specification at:
http://www.omg.org/technology/documents/modeling_spec_catalog.htm#UML

Package Diagram- A package is a general-purpose mechanism for organizing elements into groups. Or in other words, it indicates a subsystem or block of functionality within a system and is used to simplify the view of a system in order to understand it better. Figure 11-2 shows a possible package diagram for the Digital Camera System.

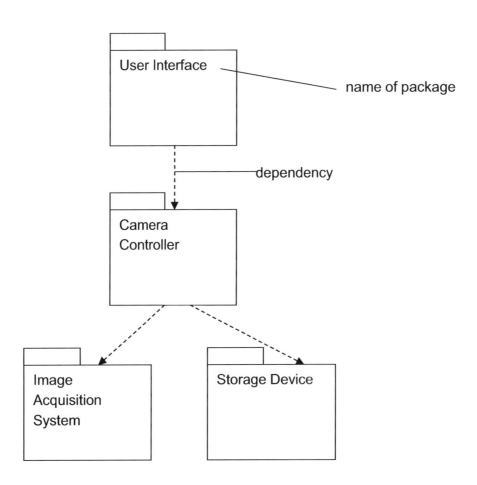

Figure 11-2 Package Diagram Example for a Simple Digital Camera System

The package diagram also shows the interactions or relationships between each package. Figure 11-2 illustrates the Camera Controller Package as a depdency on both the Image Acquisition Package and the storage Device Package. Since packages are a way of abstracting or simplfying the view of a system, each package may contain more packages, class diagrams, use cases, components, etc. until the view is at the lowest level. Well-designed packages group like elements that tend to change together. Well-structured packages are loosely coupled and highly cohesive (See Section 6.6.3 for definitions of coupling and cohesion). The graphic elements used in class diagrams to show relationships are also used in package diagrams. Package diagrams are used to show different views of a system's architecture.

Deployment Diagram - A deployment diagram shows the configuration of run time processing nodes and the components that live on each node. They are used to illustrate the static deployment view of a software system onto the hardware on which the system executes. Names of the nodes should represent the vocabulary of the hardware in the domain of the system being developed. Nodes represent a physical element and represent a computational resource having at least some memory and processing capability. Figure 11-3 shows a deployment diagram for the Digital Camera System assuming there are three processors in the system.

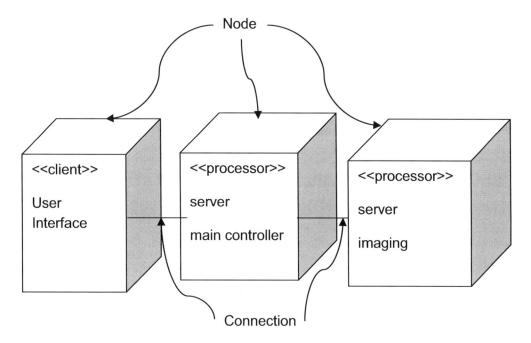

Figure 11-3 Example Deployment Diagram for a Digital Camera System

For real-time embedded software with only one processor, the deployment diagrams look like context diagrams from the system engineering world. They show the devices with which the processor board is interacting using what are called stereotypes to identify the hardware components. Client/Server and Distributed software systems have more processors on their deployment diagrams and provide the designers with an understanding of where each software component resides on the physical hardware.

The relationships between nodes are typically an association that represents a physical connection (i.e. Ethernet, cPCI bus). For more information on possible types of relationships for deployment diagrams, see the UML specification at:
http://www.omg.org/technology/documents/modeling_spec_catalog.htm#UML

11.2.3 Behavior Diagrams

Three of the eight behavior diagrams are discussed with examples in this section.

Activity Diagram - An activity diagram shows the flow from activity to activity within a system. An activity shows a set of activities, the sequential of branching flow from activity to activity, and objects that act and are acted upon. Activity digrams are used to illustrate the dynamic view of a system and are important in modeling the function of a system. Activity digrams essentially are flow charts and emphasize the flow of control among objects. Figure 11-4, shows a possible activity diagram for Taking a Picture in the digital camera system.

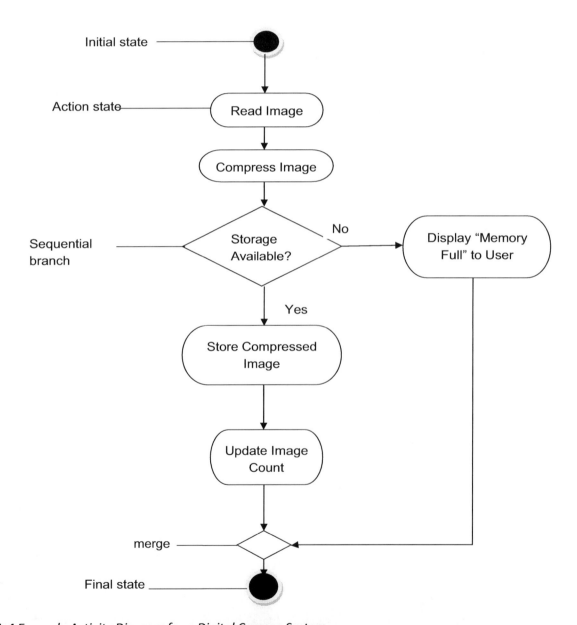

Figure 11-4 Example Activity Diagram for a Digital Camera System

Activity diagrams can also include forks to describe conditions and parallel activities. A fork is used when multiple activities are occurring at the same time. As shown on the diagram, all branches at some point are followed by a merge to indicate the end of the conditional behavior started by that branch.

Activity diagrams should be used in conjunction with other modeling techniques such as sequence diagrams and state machine diagrams. The main reason to use activity diagrams is to model the workflow behind the system being designed. Activity Diagrams are also useful for: analyzing a use case by describing what actions need to take place and when they should occur; describing a complicated sequential algorithm; and modeling applications with parallel processes[11-4]. Software engineers have also found them to be useful for scientific software development.

Sequence Diagram - A sequence diagram is an interaction diagram that emphasizes the time ordering of messages. A sequence diagram shows the objects and the messages that go between those objects at a particular instance of time. They are used to illustrate the dynamic view of a system. An example of a possible sequence diagram for the digital camera system is hown in Figure 11-5.

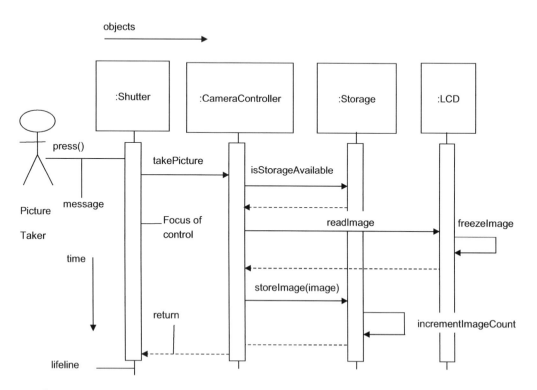

Figure 11-5 Example Sequence Diagram for a Digital Camera System

As shown in the diagram, an object has veritical lines called lifelines which represent the existence of an object over a period of time. The stick figure represents an actor, the Picture Taker. The messages become method calls when translated into executable software. The focus of control represents the period of time during which an object is performing an action directly or through a subordinate operation. Other graphic elements used in a sequence diagram can be found in the UML specification at
http://www.omg.org/technology/documents/modeling_spec_catalog.htm#UML

Sequence diagrams can be used at all levels of defining a software system – Use Cases down to a detailed software function. Although sequence diagrams are typically used to describe object-oriented software systems, they are also extremely useful as system engineering tools to design system architectures.

State Machine Diagram - A state machine is a behavior that specifies the sequence of states an object goes through during its lifetime in response to events, together with its response to those events. A state machine diagram shows the state machine, consisting of states, transitions, events, and activities. These diagrams are used to illustrate the dynamic view of a system. These are important when modeling the behavior of an interface, class, or colloboration and are useful when modeling reactive systems. Figure 11-6 shows a state machine diagram for the Shutter Object. When the button on the camera is pushed to take a picture, the Shutter object moves from the Opened state to the Closed state. Once the button is released, the Shutter Object moves back into the opened state.

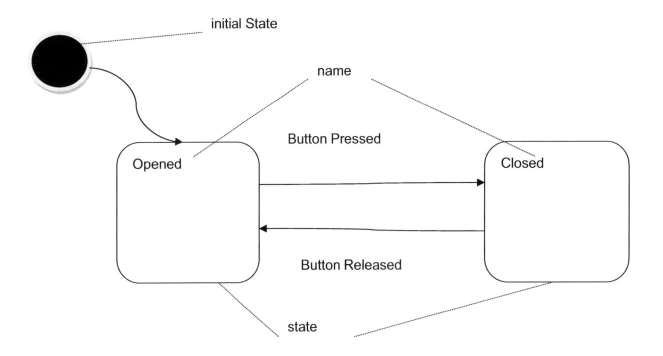

Figure 11-6 Example State Machine Diagram for a Digital Camera System

Note that a transition from one state to another could be a signal, an event, a change in some condition, or the passage of time. A state machine diagram may have a transition back to its own state. At its simplest, an object within a software system has two states, idle or running. When defining a software system, state diagrams are very beneficial for defining complex objects and how they behave.

Use Case Diagram – A Use Case Diagram is a diagram that shows a set of use cases and actors and their relationships. A Use Case is a description of a system's behavior as it responds to a request that originates from outside of that system. It provides context to what is within a system and what interacts with a system and defines the behavior of the system when it receives external stimuli (i.e. the goals of the system). A system is made up of multiple Use Cases to define the behavior of the overall system. Use Cases were first defined in 1987 at the Proceedings of OOPSLA[11-5] by Ivar Jacobson for use in Software Engineering to define functional requirements. Ivar Jacobson, et al, later published a book titled *Object-Oriented Software Engineering: A Use Case Driven Approach* in 1992[11-6] based on his experiences with Use Cases while working on large telecommunications systems. Figure 11-7 shows a Use Case Diagram for the simple Digital Camera System.

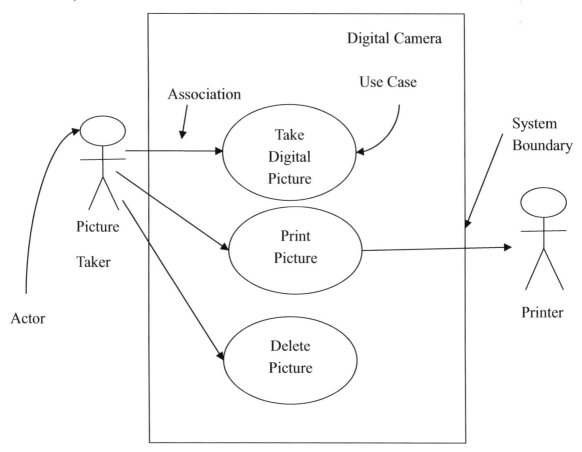

Figure 11-7 Example Use Case Diagram for a Digital Camera System

The Picture Taker and the Printer are outside the boundary of the system being defined and are called **Actors**. Take Digital Picture, Print Picture, and Delete Picture are all use cases within the system (i.e. functions the Digital Camera performs). When developing Use Cases, all Actors are described as "nouns" while the Use Cases are "verb phrases" that describe an action the system performs. The arrow on the association line indicates initiation of interaction. If either the Use Case or the Actor can initiate interaction, then there is no arrow on the relationship line. There can be two types of actors depicted on a Use Case diagram: Primary Actors which benefit from the Use Case (arrow from Actor to Use Case) and secondary actors which participate in the system but don't get benefit (arrow from Use Case to Actor).

Use Case modeling not only includes the model, but also written text to describe how the system behaves when the external Actor stimulates the system. Typically a Use Case template is defined so all needed information is captured on how the system behaves. One example of a Use Case template for capturing the behavior of a system is shown in Figure 11-8.

Use Case ID:			
Use Case Name:			
Created By:		Last Updated By:	
Date Created:		Date Last Updated:	

Actors:	
Description:	
Trigger:	
Preconditions:	1.
Postconditions:	1.
Normal Flow:	1.
Alternative Flows:	
Exceptions:	
Includes:	
Priority:	
Frequency of Use:	
Business Rules:	
Special Requirements:	
Assumptions:	
Notes and Issues:	

Figure 11-8 Example Use Case Template

When starting Use Cases, it's advisable to keep the description at the top-level at first and add details as the Use Case is better understood. Typically a Use Case is started with the normal flow of events; alternate and exception paths are added if needed. One of the strengths and a significant benefit of Use Cases is that it makes engineers think about how they want the system to behave under abnormal or failure conditions.

11.3 Creating an Executable Model

Several companies have tools available that allow for modeling systems using UML (or SysML) diagrams. These companies include EmbeddedPlus Engineering[11-7], Vitech and IBM® under their **Rational®** product suite. The tools support the modeling language semantics so the engineer can focus on creating the design of the system and not on the accuracy of the diagrams per the modeling language specifications.

Vitech supports the following UML diagram types in their **CORE**[11-8] Software:

- Activity
- Sequence
- Class
- Package
- Use Case

IBM® Rational® supports the following UML diagram types in their **Statemate**[11-9] product:

- Use Case
- Sequence
- State Machine

Using a standard development process, these tools are used from requirements down to executable software. Benefits from creating an executable model include verifying completeness and correctness of the system and bridging the gap between the systems engineering functional domain to the Object-Oriented Software Engineering domain. The models are not just done at the beginning of system definition, but evolve as the development process progresses until there is executable software. The SysML Forum, **http://www.sysmlforum.com/,** provides an overview of possible SysML tools that can be used for creating an executable model.

11.4 Benefits and Limitations of UML

Benefits 0f using UML when defining a system include:

- Standardized (by OMG Group), not proprietary
- Common language for communicating
- Explained and described in every aspect by vast amount of publications, resources, textbooks, etc.
- Can be customized and extended for specific application domain, software process, or implementation platform
- Uses object oriented design concepts
- Independent of specific programming language

SysML benefits include:

- Requirement modeling support provides the ability to assess the impact of changing requirements to a system's architecture
- Precise language, including support for constraints and parametric analysis that allows models to be analyzed and simulated, greatly improving the value of system model compared to textual system descriptions
- Open standard

Whereas UML has many benefits, it also has limitations:

- Still no specification for modeling of user interfaces
- Poor for distributed systems – no way to formally specify serialization and object persistence
- Requires training/certification
- Specification is large and takes time to understand
- Can't describe relationships between complex system composed of both hardware and software

11-4-1 Alternatives to UML(SysML) – Although SysML has become a de-facto standard not all experienced engineers find it satisfactory for their work. This is particularly true of non-software engineers with considerable experience with modeling techniques and tools aimed at their specialty. SysML tries to cover everything and that can lead to complication that causes confusion. It also means that extensive training is needed for engineers to use SysML effectively. There are alternatives, many that cover only some specific aspects of systems. Organizations new to MBSE are urged to explore UML(SysML) and other alternatives to determine if it's better for their work to put the necessary effort into becoming proficient users of the SysML, the standard, or if other tools are better suited to their work and their engineers. It requires considerable effort to choose the best tool but the cost and frustration of using a tool poorly suited to an organization's work and people makes the effort worthwhile. Some alternatives that serve to start an investigation are listed here:

- Architecture Description Language (ADL) – A family of languages; examples include ACME,(http://www-2.cs.cmu.edu/~acme/) , Rapide, (http://complexevents.com/stanford/rapide/) and UniCon, (http://www.cs.cmu.edu/~UniCon/)
- Domain Specific Modeling Language – See http://scholar.google.com/scholar?q=Domain+Specific+Modelling+Language&hl=en&as_sdt=0&as_vis=1&oi=scholart for an introduction
- Temporal modeling techniques, like LUSTRE™ (http://wiki.lustre.org/index.php/Main_Page) with associated tool SCADE Suite® (http://www.esterel-technologies.com/products/scade-suite/)
- Multi-physics functional modeling like MODELICA (https://www.modelica.org/documents) with associated simulation tools (both Open Source and commercial ones)
- AltaRica, a language designed to model both functional and dysfunctional behaviors of critical systems; useful for reliability and dependability analysis (http://altarica.labri.fr/forge/)
- Tools with application specific modules, e.g. COMSOL (http://www.comsol.com/)

- Phoenix Integration's ModelCenter®, a graphical environment for automation, integration, and design optimization (http://www.phoenix-int.com/software/phx_modelcenter.php)

(We are indebted to Jean-Luc Wippler, the current MBSE Working Group co-chair in AFIS (French Association for System Engineering, INCOSE affiliate association and other readers of our blog for suggesting these alternatives.)

11.5 Where to Find More Information on UML (SysML)

To learn more about SysML and the different diagrams, please see **http://www.omgsysml.org/INCOSE-OMGSysML-Tutorial-Final-090901.pdf**

There are many available resources on UML both in book form and on the internet. Beneficial books include:

1. *Systems Engineering with SysML/UML: Modeling, Analysis, Design* by Tim Weilkiens
2. *Model-Based Development: Applications*, by H. S. Lahman
3. *Using UML: Software Engineering with Objects and Components,* by Perdita Stevens
4. *Software Modeling and Design: UML, Use Cases, Patterns, and Software Architectures*, by Hassan Gomaa
5. *UML for Real: Design of Embedded Real-Time Systems,* by Luciano Lavagno, Grant Martin and Bran V. Selic
6. *Model-Driven Development with Executable UML (Wrox Programmer to Programmer),* by Dragan Milicev
7. *UML 2.0 in a Nutshell,* by Dan Pilone and Neil Pitman
8. *SysML for Systems Engineering (Professional Applications of Computing),* by J. Holt and S. Perry
9. *Writing Effective Use Cases*, by Alistair Coburn
10. *Software for Use: A Practical Guide to the Models and Methods of Usage-Centered Design* , by Larry L. Constantine and Lucy A. D. Lockwood
11. *Use Case Modeling ,* by Kurt Bittner and Ian Spence
12. *Scenarios, Stories, Use Cases: Through the Systems Development Life-Cycle*, by Ian Alexander, Neil Maiden
13. *Use Case Driven Object Modeling With UML: Theory And Practice*, by Doug Rosenberg, Matt Stephens

Beneficial web sites include:

1. Unified Modeling Language Resource Page, **http://www.uml.org/**
2. Systems Modeling Language Resource Page, **http://www.omgsysml.org/**
3. SysML Forum, **http://www.sysmlforum.com/**
4. Embarcadero Developer Network, **http://edn.embarcadero.com/article/31863**
5. Unified Modeling Language Tutorial, **http://atlas.kennesaw.edu/~dbraun/csis4650/A&D/UML_tutorial/**

6. OMG Systems Modeling Language Tutorial, **http://www.uml-sysml.org/documentation/sysml-tutorial-incose-2.2mo**
7. An Introduction to Systems Engineering with Use Cases, by Ian Alexander and Thomas Zink, **http://easyweb.easynet.co.uk/~iany/consultancy/use_cases/use_cases.htm**
8. Visual Paradigm, **http://www.visual-para-digm.com/product/vpuml/provides/umlmodeling.jsp?src=google&kw=use%20cases&mt=p&net=s&plc=&gclid=CJSQzqzNkqgCFYi8KgodmmeJCw**
9. SmartDraw, **http://www.smartdraw.com/specials/ppc/softdesign.htm?id=10514&gclid=CNSquMTNkqgCFcW5KgodrHR2Dg**

Exercises:

Rather than work any exercises based on the material presented in this chapter read the recommended material in the web sites number 6 & 7 in the list above.

12 INTEGRATING MODERN METHODS FOR FASTER SYSTEMS ENGINEERING

12.0 Introduction

In chapter 2 it was explained that the best model for system development is the "craftsman" model that was widely used before systems became so complex that a single chief engineer could no longer understand a system in sufficient detail to control all aspects of design. System engineers, design engineers and other specialty engineers became necessary to handle the complexity of modern systems. Although this new approach has enabled the development of very complex modern systems it takes much longer to develop a system now than it used to take when a chief engineer and his/her team could develop a new system in a few months.

One objective of this book is to introduce new methods that enable the systems engineering work on system development to be accomplished faster and more accurately. This book has an emphasis on systems engineering fundamentals, as described in the DoD SEF and the NASA SE handbook, and readers will note that it takes time and discipline to follow these fundamental processes. Complex systems cannot be developed cost effectively by shortcutting the systems engineering fundamentals; what is needed is faster and more accurate methods for executing these fundamentals. Accuracy is required because any errors in systems documentation results in costly "find and fix" efforts later in design or in integration and test. Several methodologies for ensuring accuracy have been discussed including using graphical models in place of text as much as possible, employing redundant tools for consistency checks, using modeling and simulation to support requirements analysis as well as design and checking work at the three levels of worker checking his/her work, peer reviews and design reviews.

In chapter 5 **Pattern Based Systems Engineering** was introduced, which when properly implemented, can dramatically reduce the time to produce much of the top level systems engineering documentation and at the same time increase the accuracy of requirements definition. Similarly, using validated system performance models and simulations throughout the development cycle aids in reducing development time and increases the accuracy of requirements and design concepts and the robustness of systems.

The objective of this chapter is to describe methods for reducing **information latency** and then to show how integrating modern methods can achieve greatly reduced time for systems engineering work without sacrificing any process fundamentals critical to the accuracy of this work. Information latency is the time between when information is generated and the time it is available to others who are depending on the information for the next steps in their work. Information latency was increased with the evolution from the craftsman model for product development to models with systems engineers and other specialty engineers; this is the primary reason modern systems take so long in development. Reducing information latency to levels near what it was for the craftsman model is a necessary step in achieving faster system development cycles.

12.1 Integrated Concurrent Engineering

In the 1990s a method emerged for reducing information latency for system development teams. This method is similar to methods used previously when teams of workers were brought together in a common work area to collaborate to quickly accomplish some project. Many organizations in the aerospace and defense industry use special work areas to colocate the people writing and publishing proposals, which are often highly time constrained projects. The use of proposal preparation rooms for the personal assigned to working on proposal preparation results in highly productive teams for the limited times involved in typical proposal efforts. A major part of the increased productivity is due to the reduction in information latency achieved by having workers so close they can ask questions of one another and get immediate answers. If teams try to maintain such intense work over long periods productivity tapers off due to workers being unable to maintain the long hours and intense work without burnout.

The methods that evolved in the 1990s achieve the reduction in information latency and the associated productivity gains of the colocated teams and permit teams to work effectively for long periods without burnout. These methods became possible by exploiting new technology as well as new work management methods.

The availability of inexpensive large screen display projectors, n to one video switches and inter/intra nets makes it cost effective to set up special work rooms where teams of 10 to 25 knowledge workers can gather with their laptops or pc's and software tools. These teams can simultaneously work and share the work results with the entire team on the large screen displays as fast as the results are available. Many organizations now use such facilities for teams to gather for intense work and information sharing periods of three to four hours two or three times weekly. These sessions must be well planned and workers must come prepared to work and share results in real time. Planning, documenting work and time consuming analysis tasks are performed in between the sessions in the special work rooms. This approach is called by a number of names but **Integrated Concurrent Engineering** (ICE) is a common name. This approach is effective because it reduces information latency from minutes to seconds or hours to minutes.

ICE is proven to reduce cost and schedule of complex projects by factors of three to ten [12-1, 12-2]. Neff & Presley [12-3] reported that the Jet Propulsion Laboratory initially achieved an average of over 80% reduction in project costs and significantly improved the quality and speed of work. With more experience a 92% reduction in design time and a 66% reduction in cost was reported. Designs produced using ICE are of higher quality because they examine each option in greater detail earlier in the design process by sharing thousands of design variables in real time. Approaches that are proven to reduce cost and schedule by factors of three to ten and increase quality at the same time should not be dismissed by organizations that wish to remain competitive.

The benefits of ICE are better understood by examining the work space and the work process in more detail. There is no single best work space design or work process; each organization tailors both to their needs and their business processes. Examples presented here are guidelines for understanding ICE and not necessarily the best for any specific organization.

12.1.1 The ICE Design Command Center- A schematic diagram of a small ICE work area is shown in Figure 12-1. The room has large screen displays located where they are visible to everyone in the room. Several displays are used so that several different types of information can be displayed simultaneously. Each skill cluster has workers with common specialties and each worker has computer equipment and the design, modeling and simulation tools associated with his/her specialty. Alternatively each cluster can be an IPT responsible for a segment of the system design. Each of the computers is connected to one of the large screen displays via a video switch so that the results of analysis, modeling or simulation can be shared with everyone in the room on one of the large screen displays. The facilitator, typically the lead systems engineer for the systems engineering phase of development, is responsible for maintaining the design baseline visible to all at all times and to lead the team through a preplanned sequence of analysis tasks that facilitate design decisions in real time.

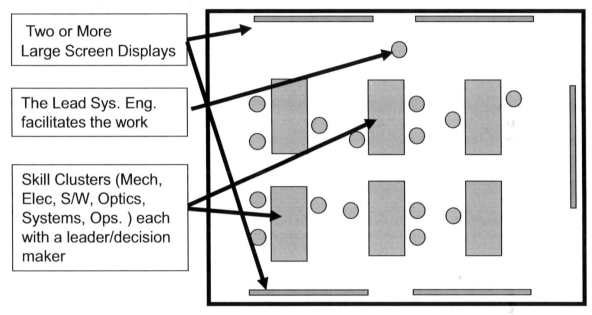

Figure 12-1 A diagram of a small ICE work are showing six specialty skill clusters, five large screen displays and a facilitator/leader.

Many developments take place in multiple locations and it isn't practical to bring teams from different locations together in a common work area for each ICE session. Modern technology enables teams working is several separately located ICE facilities to communicate with each other and to see each other's work in real time; thereby achieving the benefits of ICE even though the teams are in different locations. Technology allows virtual colocation. Using ICE with a team that is fractionated into multiple locations and even diverse cultures has the benefit of facilitating the interactions necessary to the success of any system development.

12.1.2 The ICE Concept of Operations - Integrated Concurrent Engineering is a repeating series of planning sessions followed by team work sessions, followed by documentation and follow-up analysis in parallel with the planning for the next series of team work sessions. The times for each of the components of the ICE cycle are dependent on the type and complexity of the system being developed. Example times are given here to explain the concept of operations. Development teams are likely to find adopting this concept of operations to their systems development requires adjustments. A typical approach is illustrated in Figure 12-2 where a series of three plan/meet/document sessions are shown and each of the meet or design sessions is comprised of three intense team sessions separated by a day or two. Individual design sessions may last from two to four hours.

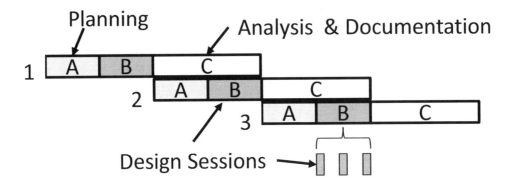

Figure 12-2 An example ICE concept of operations where A is planning, B is series of team work sessions and C is documentation and analysis.

The planning, indicated by A in Figure 12-2, is done by team leaders and might take a week to plan a series of three intense work sessions, indicated by B, over another week period. The series of work sessions is followed by perhaps two weeks of documenting work done in the design sessions and carrying out analyses that takes too much time to be done in design sessions. In the example shown in figure 12-2 nine intense design sessions are planned, executed and documented in a an eight week period. Note that since the design sessions are the only activities that require the ICE design command center such a center can support three or four ICE projects or separate IPTs of a large project concurrently.

12.2 ICE for Small Teams

The ICE approach described in Section 12.1.1 and 12.1.2 applies to teams of 15 to several hundred people; assuming the large teams are organized into smaller IPTs of 10 to 25 people. The design command centers can be shared by many individual teams on a development project because each team uses the center for only a half day at a time and for only three to ten days a month typically. Some system developments can be accomplished with smaller teams of five to ten people. Whereas small teams can also use the same design command center and concept of operations as larger teams an alternative approach may be even more efficient.

Work spaces for most organizations use individual cubicles or cubicles shared by two or three people. Most of these work spaces are modular and can easily be reconfigured. For example suppose a project has six or seven workers each in his/her cubicle. Typically, workers are assigned cubicles without consideration of where others working on the same projects are located. Much of the communication takes place via emails or periodic meetings in a conference area. Figure 12-3 shows how a space of eight cubicles can be rearranged to colocate seven workers and a conference table. Collocating workers as shown in Figure 12-3 enables continuous face to face interactions to replace emails and periodic meetings in conference rooms. Research has shown that problems are solved much faster by groups communicating face to face compared to groups communicating via email. That is to be expected because the information latency in face to face communications is almost instantaneous whereas it is many seconds or even hours with email.

It increases productivity to have two workers with related skills close enough together that they can see each other's computer screens and discuss what is on the screen without moving from their work positions. Examples include mechanical and thermal engineers or mechanical engineers and designers skilled in mechanical CAD tools that are supporting the engineers.

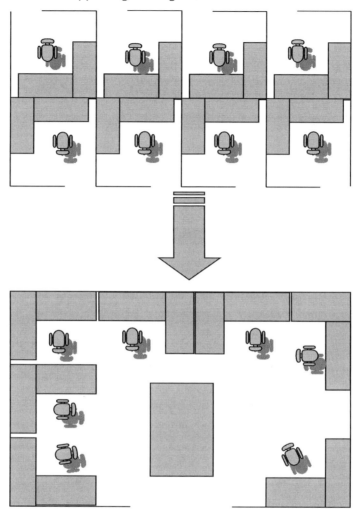

Figure 12-3 Individual cubicles can be rearranged to form a space allowing a team of four to eight to see each other's computer screens and exchange information without moving from their work station.

If the team leader is collocated with the rest of the team so that he/she can facilitate an ICE process then dramatic reductions in project cost and design time should be realized just as it is for larger teams using ICE. A caution is that team dynamics are more important for collocated teams than for teams in individual cubicles. Teams must be comprised of individuals who work well together or else productivity suffers. Workers who perform better as individual contributors are likely better left in their own cubicle. It is also advisable to provide training so that the workers understand why they are being asked to give up the privacy of individual cubicles.

12.3 Return to Chief Designer Model

Implementing ICE allows system development teams to function similarly to the model of chief designer and draftsman/assistant team popular before the emergence of modern complex systems in the 1960s.

The large screen displays in a design command center and the supporting analysis models and simulations bring design information to the lead systems engineer with very little information latency. The lead systems engineer in a design session can interact with the design team just as a chief designer interacted with the draftsman/assistants in former times. This may be as near to the efficiency of the craftsman model as can be expected for the development of complex systems. Lead systems engineers can be empowered to function as chief designers for the systems engineering work in a mature ICE environment supported by comprehensive analysis, modeling and simulation tools. The lead systems engineer can be empowered to function as the chief designer for the entire development cycle if supported by specialist chief designers who are responsible for the electrical design, the mechanical design, etc.

Implementing ICE with an overall chief designer and supporting specialty chief designers for each IPT allows interleaving IPT design sessions with SEIT design sessions so that the desired iteration between levels of design and the coordination between IPTs necessary to maintain balance in the design can be achieved and the schedule for the development is likely to be significantly reduced.

The actual times it takes for the planning and for the documentation and analysis periods are highly dependent on the sophistication of the tools used by the design team. If pattern based systems engineering is used and if the team's modeling and simulation tools are extensive and mature then it may be possible to integrate the planning and the documentation/analysis periods with the design work to achieve a continuous series of three to four hour intense design sessions in the design command center followed by a day or two of planning/documentation/analysis, followed by another design session. Alternatively, the team may be organized with design specialists and documentation specialists. The design specialists conduct analysis, modeling and simulations to determine design parameters. The documentation specialists capture the design parameters and product the necessary specifications, drawings and CDRLs while the design specialists are generating the next layer of design parameters.

12.4 Integrating Modern Methods

The 21st century brought new constraints to system development:

- Customers and global competition are demanding faster and cheaper system development
- Skilled engineers are retiring faster than replacements are experienced enough to replace them
- Development teams are spread across multiple sites and multiple organizations.

This new century has also brought new tools for system development:

- Fast internet and intranet connections provide real time communication across multiple sites
- Relatively cheap but powerful computers and network communication tools
- Model based and Pattern Based Systems Engineering processes
- Powerful CAD tools
- Maturing integrated design and design documentation processes
- Some integrated design and manufacturing tools
- Potential for end to end documentation management

The question for systems engineers is how to use the new tools to relieve the new constraints.

One answer to this question is to integrate the methods described in this and previous chapters with disciplined execution of the traditional fundamentals of the systems engineering process.

Figure 12-4 illustrates methods that can be synergistically integrated to achieve reductions in design time of factors of three to ten and cost reduction by factors of two to three. These benefits are not achieved instantly. Training is needed for teams to use these methods effectively. Investment is necessary to achieve the best results of PBSE and to push patterns down from the system level to subsystem and assembly levels. Ongoing investment is necessary to maintain the modeling, simulation, software development and CAD/CAM tools required to remain competitive. Document generation and document management tools are likely to require investments and training to effectively reduce engineering effort. Finally it must be recognized that systems engineering is going to continually evolve by inventing new processes and tools and by introducing new methods and tools for executing current processes.

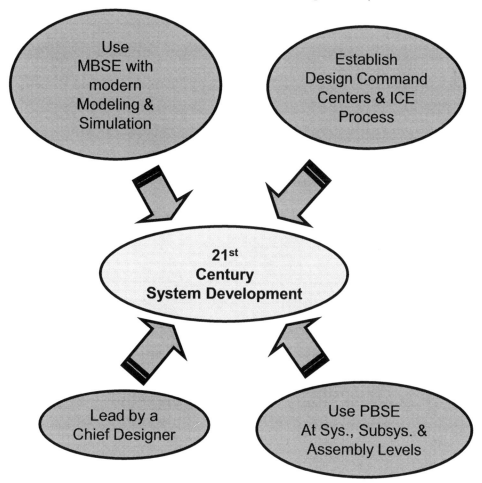

Figure 12-4 The methods described in this book can be integrated to provide a robust approach to system development that can achieve dramatic reductions in cost and design time.

The rapid introduction of new tools and processes in the past two decades have increased the fraction of a systems engineer's time that must be spent in training and self-study in order to maintain required skills. This is likely to continue. The increases in complexity of new systems are also likely to continue and these complexity increases may require more sophisticated systems engineering processes than available today. Hopefully new methods and tools will be developed that can handle increased system complexity and the increases in productivity from using new methods are enough to make time available for the training and self-study systems engineers will need.

Exercises:

1. Review a current development project in your organization and identify ways in which information needed by engineers is delayed in getting to the engineers after it has been generated.

2. Consider how the methods shown in Figure 12-4 could reduce information latency in your organization.

3. Consider a typical system development in your organization. How much time would be needed to plan an ICE design session for such a project? How much time might be needed to document the results of a design session? Construct a diagram similar to Figure 12-2 for an ICE process tailored to system development in your organization. What are the barriers to implementing ICE in your organization?

REFERENCES

1-1 The Department of Defense Systems Management College publication *Systems Engineering Fundamentals,* Supplementary text prepared by the Defense Acquisition University Press Fort Belvoir, Virginia, Jan. 2001 Download in PDF at
http://spacese.spacegrant.org/SEModules/Reference%20Docs/DAU_SE_Fundamentals.pdf

1-2 NASA *Systems Engineering Handbook* Sp-2007-6105 Rev 1 Download in PDF at
http://education.ksc.nasa.gov/esmdspacegrant/Documents/NASA%20SP-2007-6105%20Rev%201%20Final%2031Dec2007.pdf

1-3 IEEE Std 1220-1998 *Standard for Application and Management of the Systems Engineering Process* Download in PDF at **http://teal.gmu.edu/ececourses/tcom690/spring2004/IEEE-Standard-for-SE.pdf**

1-4 *INCOSE* (International Council on Systems Engineering) *Systems Engineering Handbook*
http://www.incose.org

1-5 GPR 7120.5A *Systems Engineering* (GPR is NASA Goddard Procedural Requirements) Download in Word at **http://www.everyspec.com/NASA/NASA+-+GSFC/GSFC-GPR/download.php?**spec=GPR_7120_5A_Systems_Engineering.194.doc

1-6 Technical papers published by the Vitech Corporation are available at
http://www.vitechcorp.com/support/papers.php

1-7 The Systematica™ Methodology of ICTT System Sciences See
http://www.ictt.com/Systematica.html

1-8 *"Observation, Theory, and Simulation of Integrated Concurrent Engineering: Grounded Theoretical Factors that Enable Radical Project Acceleration"* by John Chachere, John Kunz, and Raymond Levitt available at **http://cife.stanford.edu/online.publications/WP087.pdf**

1-9 International Standard ISO/IEC 15288 Systems and software engineering —System life cycle processes, Second edition 2008-02-0 Available at
http://web.mit.edu/deweck/Public/16_842_SE/Readings/1c-ISO%2015288%202008.pdf

1-10 *The Manager's Guide for Effective Leadership* by Joe Jenney, AuthorHouse, 2009

1-11 *Realizing maximum Product Revenue Utilizing a Product Revenue Model* by Martin Coe, 18th International Conference on Systems Engineering, Las Vegas, NV, August 2005

1-12 Department of Defense Instruction (DoDI) 5000.02, *Operation of the Defense Acquisition System* The December 8, 2008 version is available at **http://www.dtic.mil/whs/directives/corres/pdf/500002p.pd**

2-1 *Total Quality Development-A Step-By-Step Guide to World-Class Concurrent Engineering* by Don Clausing, ASME Press, 1994

2-2 *Quality Engineering Using Robust Design* by Madhav Phadke, PTR Prentice Hall, 1989.

2-3 *Strategic Design: A Guide to Managing Concurrent Engineering* by Bart Huthwaite, 1994

3-1 *Systems Architecting: Creating & Building Complex Systems* by Eberhardt Rectin, Prentice Hall; 1st edition, 1990

4-1 See **http://en.wikipedia.org/wiki/N2_chart**

4-2 *Production and Operations Management,* by James B. Dilworth, Random House, 1986

4-3 **http://herdingcats.typepad.com/my_weblog/2006/02/deterministic_v.html** A web site that treats probabilistic scheduling.

4-4 *Critical Chain* by Eliyahu M. Goldratt North River Press; Illustrated edition (April 1997)

5-1 *Pattern-Based Systems Engineering: An Extension of Model-Based SE* by Bill Schindel, ICTT, INCOSE 2005 TIES 4

http://www.incose-coa.org/29NOV07/ICTT%20TIES%20Tutorial%20V1.2.2.pdf

5-2 *Requirements Statements Are Transfer Functions: An Insight from Model-Based Systems Engineering,* by W.D. Schindel, ICTT, Inc., and System Sciences, LLC[5-2]. Presented at the 15th Annual International Symposium, INCOSE 2005, 10-14 July 2005

6-1 The Vee diagram. Forsberg, Kevin, and Harold Mooz, "*The Relationship of System Engineering to the Project Cycle",* Center for Systems Management, 5333 Betsy Ross Dr., Santa Clara, CA 95054; also available in *A Commitment to Success,* Proceedings of the first annual meeting of the National Council for Systems Engineering and the 12th annual meeting of the American Society for Engineering Management, Chattanooga, TN, 20-23 October 1991.

6-2 *Cognitive Fit Applied to Systems Engineering Models* by Larry Doyle and Michael Pennotti, Conference on Systems Engineering Research, 2004

8-1 See **http://www.galorath.com/**

8-2 See **http://www.pricesystems.com/**

9-1 See **http://www.dau.mil/conferences/presentations/2006_PEO_SYSCOM/tue/A2-Tues-Stuckey.pdf**

9-2 See for example **http://esto.nasa.gov/files/TRL_definitions.pdf**

10-1 *The Manager's Guide for Effective Leadership* by Joe Jenney, page 205, AuthorHouse, 2009

10-2 *Risk Management Guide for DOD Acquisition,* Sixth Edition (Version 1.0), Department of Defense, August 2006 **http://www.dau.mil/pubs/gdbks/risk_management.asp**

11-1 *Model-based Systems Engineering (MBSE) Initiative*, by Mark Sampson and Sanford Friedenthal, Presented at the Opening Plenary of the International Workshop, Phoenix, AZ, 29 January 2011; **http://www.omgwiki.org/MBSE/lib/exe/fetch.php?media=mbse:mbse_iw_2011_intro-b.pdf**

11-2 *Foundational Concepts For Model Driven System Design* by Loyd Baker, Paul Clemente, Bob Cohen, Larry Permenter, Byron Purves, and Pete Salmon, INCOSE Model Driven System Design Interest Group

11-3 *The Unified Modeling Language User Guide*, by Grady Booch, James Rumbaugh, and Ivar Jacobson, Addison-Wesley Professional; 2 edition, May 29, 2005

11-4 Unified Modeling Language Tutorial, **http://atlas.kennesaw.edu/~dbraun/csis4650/A&D/UML_tutorial/activity.htm**

11-5 *Object-Oriented Development in an Industrial Environment*, Ivar Jacobson, Proceedings of OOPSLA´87, SIGPLAN Notices, Vol. 22, No. 12, pages 183-191, 1987

11-6 *Object-Oriented Software Engineering: A Use Case Driven Approach,* by Ivar Jacobson, Magnus Christerson, Patrik Jonsson, and Gunnar Övergaard, Addison-Wesley, Wokingham, England, 1992.

11-7 Embedded Plus Engineering Web Site**, http://www.embeddedplus.com/SysML.php**

11-8 CORE Software Web Site, Vitech, **http://www.vitechcorp.com/products/index.html**

11-9 Rational Statemate Web Site, **http://www-01.ibm.com/software/awdtools/statemate/**

12-1 *The Integrated Concurrent Enterprise* by David B. Stagney, MIT Department of Aeronautics and Astronautics, Sloan School of Management, August 13, 2003

12-2 *Observation, Theory, and Simulation of Integrated Concurrent Engineering* by John Chachere, John Kunz, and Raymond Levitt, Center For Integrated Facility Engineering, Working Paper #WP087, Stanford University, August 2004 **http://cife.stanford.edu/online.publications/WP087.pdf**.

12-3 *Implementing a Collaborative Conceptual Design System – The Human Element is the Most Powerful Part of the System* by Jon Neff and Stephen P Presley, IEEE, 2000.

ACRONYMS

AV	All View
BOE	Basis of Estimate
CBE	Current Best Estimate
CCPM	Critical Chain Project Management
CDD	Capability Development Document
CDRL	Contract Data Requirments List
Con-Ops	Concept of Operations
COTS	Commercial Off the Shelf
cPCI	Conventional Peripheral Component Interconnect
CPM	Critical Path Method
CWBS	Contract Work Breakdown Structure
DARPA	Defense Advanced Research Projects Agency
DMR	Data Management Repository
DoD	Department of Defense
DoDAF	DoD Architecture Framework
DoDI	Department of Defense Instruction
DOE	Design of Experiments
EMI	Electromagnetic Interference
EOL	End of Life
FA/A	Functional Analysis/Allocation
FFBD	Functional Flow Block Diagram
FMEA	Failure Modes and Effects Analysis

IED	Improvised Explosive Devices
IMP	Intergrated Management Plan
IMS	Integrated Master Schedule
INCOSE	International Council on Systems Engineering
IPD	Integrated Product Development
IPPD	Integrated Product and Process Development
IPT	Integrated Product Team
IR&D	Internal Research and Development
KPP	Key Performance Parameters
KSA	Key System Attributes
LRU	Line Replaceable Unit
MBSE	Model Based Systems Engineering
OMG	Object Management Group
OOPSLA	Object-Oriented Programming Systems, Languages, and Applications
OV-1	Operational View -1
PBS	Product Breakdown Structure
PBSE	Pattern Based Systems Engineering
PERT	Program Evaluation and Review Technique
PLC	Product Life Cycle
PLM	Product Life Cycle Management
PPM	Pre-Planning Matirx
QFD	Quality Function Deployment
QRC	Quick Reaction Capability
RF	Radio Frequency
RFP	Request for Proposal
RFS	Release for Sale
RMA	Reliability/Maintainability/Availability
RMS	Root Mean Square
SBS	System Breakdown Structure

SCD	Source Control Document	
SDD	System Design Document	
SDR	System Design Review	
SE	Systems Engineering	
SEDS	Systems Engineering Detail Schedule	
SEF	Systems Engineering Fundamentals	
SEIT	System Engineering and Integration Team	
SEMP	Systems Engineering Management Plan	
SEMS	Systems Engineering Master Schedule	
SEP	Systems Engineering Plan	
SITP	System Integration and Test Plan	
SOO	Statement of Objectives	
SRD	System Requirements Document	
SRR	System Requirements Review	
SV-x	Systems and Services View	
TBD	To Be Determined	
TBR	To Be Reviewed	
TPM	Technical Performance Measure	
TRR	Test Readiness Review	
TV-x	Technical Standards View	
UML	Unified Modeling Language	
US	United States	
VOC	Voice of the Customer	
WBS	Work Breakdown Structure	
WCA	Worse Case Analysis	
WFD	Work Flow Diagrams	
WSS	Weapon Systems Specification	

INDEX

activity diagram 200, 203-205

actors 206-208

allocated requirements 85, 127

AltaRica 210

American Supplier Institute 139

analytic hierarchy process 165

Architecture Description Language 210

balanced design 18, 19, 25, 78, 111

baseline design 83, 127, 129, 160, 161, 166, 174

basis of estimate 105, 106

behavior diagrams 200, 203

benchmark 162, 163

best value design 168, 170, 174

calendar schedule 41

capability development document 13

cardinal requirements 103, 134, 161, 189, 190

chief designer model 217

chief engineer 16, 18, 19, 24, 31, 34, 35, 213

class diagram 200-202

Clausing 20, 22, 97

cohesion 129, 130, 202

commercial product development 7, 9, 10

communication diagram 200

competitive assessment 148, 149, 152, 153

compliance matrix 180, 185, 186

component diagram 200

composite structure diagram 200

computer aided design 1, 198

COMSOL 210

concept design 126-130, 160

concept of operations 84, 89, 94-97, 116, 215, 216

concurrent engineering 4, 19, 20, 23, 25, 33, 53

configuration management 39, 54, 84, 127

configuration management plan 54

connectivity 129, 130

context diagram 84, 90-92, 94-98, 117, 203

contingency 103-105, 183

contract database 60

contract work breakdown structure 53

control factors 95, 96

core team 33, 34, 36

CORE© 14, 61, 209

correlation matrix 143, 146, 147, 152

cost benefit analysis 165

cost modeling 160, 175

coupling 129, 156, 202

CRADLE 61, 134

craftsman model 24, 25, 31, 35, 179, 218, 219

critical chain project management 47

critical path analysis 50

critical path method 46

current best estimate 103, 104

data dictionary 84, 92, 99, 120

data management repository 59

decision management 1, 23, 30, 69, 72, 75, 127

decision matrix 69, 163

decision reuse 72

decision trees 161, 165

derived requirements 85

Descartes 67

design command center 214, 216, 218

design database 59, 83

design iteration 20, 21, 23, 190, 196

design margin 16, 18, 103

design model 199

design requirements 85, 134, 138, 139, 174

design synthesis 78, 80, 85, 110-112, 126, 127, 130, 160, 177, 178,

design trade matrices 161, 163, 175

design validation 12

design verification 12, 84

detail schedule 41-43, 47, 53

detailed design 23, 59, 126

development cycle 3, 6, 7, 19, 20, 27, 28, 31, 76, 104, 133, 213, 214, 218

document tree 57, 74, 83

DoDAF 14, 15

domain diagrams 89, 114

Domain Specific Modeling Language 210

DOORS™ 13

enablers 82

environmental models 166

executable model 198, 199, 209

executable schedule 42

failure analysis 84, 111, 179

FMEA 179

functional analysis 22, 69, 80, 81, 94, 96, 99, 110, 112, 121, 122, 130, 156, 160, 178, 185, 198

functional architecture 110, 112, 121, 122, 126, 127, 130, 152, 156, 158, 178

functional decomposition 99, 115

functional flow block diagram 74, 112

functional interfaces 117, 120

functional requirements 85, 92, 95, 98, 121, 200

functional to physical allocation matrix 130, 131

functional trees 116

functional view 63, 84, 85, 110, 112, 121, 132

Gantt chart 48, 49

house of quality 137, 139, 141, 152

Huthwaite 20, 108

ideal function 95-97

ility diagram 108

importance rating 148-151, 153-155

INCOSE 4, 5, 81, 82, 128, 202, 204, 216, 217

influence diagrams 165

information architecture 39, 55, 57, 59, 74, 83, 126

information latency 1, 8, 17-19, 24, 213, 214, 218, 220

integrated concurrent engineering 4, 25, 219, 220

integrated management plan 38, 39, 74

integrated master schedule 41, 42, 47

integrated product development 1, 20, 66

integration definition for function modeling 112

interaction overview diagram 200

interaction tables 200

interface control document 14, 74

Kano diagram 101, 102, 178

key performance parameters 13, 86, 106

key system attributes 13

lifecycle cost 9, 10, 11, 12

logical architecture diagram 74

logical interface 69, 90-92, 117

logical subsystem 69

LUSTRE™ 210

margin depletion line 104, 106

master file directory 61

master schedule 41-43, 47, 48, 53, 55

matrix diagram 162

measures of effectiveness 82, 85

mechanical block diagram 171

mission assurance plans 74, 126

mode transition diagram 101, 121

model based systems engineering 1, 4, 5, 7, 61, 197, 202, 230

ModelCenter® 211

MODELICA 210

modeling and simulation plan 39, 63, 175

modes diagram 74

modular designs 129

modularity 127, 130, 132

monte carlo simulations 47

multi-criteria decision analysis 165

network schedule 41, 42, 27

noise factors 95-97

N-squared diagram 43, 45-48, 56, 112, 120, 132, 161

N-squared interface diagram 74

object diagram 200

object management group 199

object oriented design 209

object-oriented model 113, 114

operational view 14, 85, 86, 126, 137

opportunity management 187

package diagram 200, 202

parameter diagram 74, 95, 165

parameter model 166

pattern based systems engineering 4, 7, 23, 67, 68, 179, 213, 218

pattern diagram 67, 68, 73-75, 90, 91, 114

P-diagram 95-97, 102

peer reviews 8, 22, 25, 178-180, 213

performance model 167, 169, 170, 175, 198, 213

performance requirements 81, 85, 98, 110, 111, 124, 156, 158

performance verification 165, 166, 179

PERT chart 48

Phadke 20, 22, 97

physical architecture 14, 59, 69, 127, 130, 156

physical block diagram 127, 128

physical model 68, 199

physical partitioning 111, 130

physical view 61, 63, 84, 85, 87, 88, 127, 132

planning database 60, 61

preliminary design 126, 127

pre-planning matrix 153

probabilistic scheduling 47, 51, 66

process-oriented model 112

product assurance 35, 39, 40, 165

product assurance plan 39, 40

product breakdown structure 42

product development cycle 3, 19, 20, 27, 28, 31, 76

product life cycle management 3

program database 60

program evaluation and review technique 46

progressive freeze 22, 23, 25, 29, 30, 37, 196

Pugh concept selection 22, 161-163, 175

Pugh diagrams 109

quality function deployment 4, 7, 22, 25, 98, 102, 133-135, 158, 189

quality table 139

quick reaction capability 13

Rational® 14, 209

relationship matrix 143, 146, 150, 152

requirements allocation sheet 112, 115

requirements analysis 69-72, 81, 84-87, 95, 96, 109-111, 115, 122, 177-179, 185, 189, 213

requirements database 60, 122, 123, 174

requirements reuse 69, 71

requirements traceability 71, 184

risk definition 187

risk management 4, 11, 12, 22, 47, 48, 54, 165, 179, 187-191, 194-196

risk management plan 54

risk mitigation 48, 120, 182, 183, 189, 190, 194, 195

risk register 40, 84, 190, 194-196

risk summary grid 190-194

robust design 20, 23, 97, 178

rolling wave scheduling 43, 47, 48

sale point 153

scenarios 59, 85, 89, 94, 95, 115, 211

Schindel 68, 69

SEIT 34-37, 174, 218

sequence diagram 200, 205, 206

signal flow block diagram 127

software engineering 4, 12, 207, 209, 211

source control documents 65

specification tree 57, 59, 132

spider diagram 108

spiral development 22, 23, 196

state diagram 99, 207

state machine diagram 200, 206, 207

Statemate 209

statement of objectives 12

statistical process control 139

structure diagrams 199, 200

sub modes diagram 74

subsystem specifications 62, 125, 134

SysML™ 199

system analysis 12, 83, 124

system architecting 12, 85

system architecture 30, 109, 127, 206

system block diagram 74

system breakdown structure 40, 42

system breakdown structure 40, 42

system complexity 1, 130, 158

system design document 38, 59, 83, 175

system design review 30, 175

system engineering and integration team 34

system functional review 98

system hierarchy 24, 80, 83, 126, 129, 131, 161, 184, 185

system integration 12, 56, 64, 84, 175, 179, 180, 181, 183, 185

system integration and test plan 62, 180, 181

system performance simulation 63, 167

system requirements document 13

system requirements review 98, 180

system specification 14, 74, 83, 127, 134, 171, 184

system test 7, 64, 69, 171, 189, 190, 191

system test methodology plan 184

system test plan 184

system validation 56, 177

system verification 56, 81, 166, 178

Systematica™ 4

systems engineering detail schedule 41

systems engineering management plan 14, 38, 52

systems engineering master schedule 41

systems engineering plan 14

Taguchi 7, 20, 22, 97, 178, 179

Taguchi Design of Experiment 20, 22, 97, 178, 179

technical performance measures 55, 85

technical performance metrics 101, 102

test architecture definition 62, 180, 183, 184

test data analysis 166, 180, 185

test data analysis plans 62, 180

test plans and procedures 62, 184, 185

thermal block diagrams 170

time blocked task list 48, 49

time line analysis 112

timing diagram 200

trade study methodology 161

tree chart 106, 107

Unified Modeling Language™ 199

use case 81, 84, 198, 200, 205, 207, 208, 211

use case diagram 200, 207

VEE diagram 76, 77

verification matrix 179-182, 185, 186

verification matrix 62, 179-182, 185, 186

voice of the customer 12, 82, 133, 135, 140, 141

weekly task schedule 42

work breakdown structure 12, 36, 40, 42, 43, 53

work flow diagrams 43, 47, 48

worse case analysis 179

Yourdan 68

ABOUT THE AUTHORS

Dr. Joe Jenney is a retired executive with twenty years of experience managing, training and consulting in systems engineering. He has experience in a variety of organizations including aerospace, defense, research and development, federal government, banking, consulting and civic organizations. He learned the principles and practices of modern systems engineering from working with some of the finest systems engineers in the aerospace and defense industry, from developing and implementing training programs in systems engineering and by practicing the methods described in this book. He is also the author of the book *The Manager's Guide for Effective Leadership.*

Michael Gangl is a senior systems engineer with over 27 years' experience developing sensors for the US Department of Defense and Department of Commerce. He has had the pleasure of collaborating with some of the co-authors for both systems engineering and training. An enjoyable SE example was developing system requirements from higher level customer needs and evaluating their utility using modeling and simulation. This was performed for a space-based weather satellite. A second unique example was incorporating systems engineering processes to prioritize intelligence, surveillance and reconnaissance (ISR) solutions for military intelligence. The tools and techniques applied by systems engineers have no limits to where and when they may be applied.

Rick Kwolek is a systems engineer with 32 years' experience in the engineering field with 20 years specifically with the systems engineering discipline. He spent his entire career working in the defense industry where products must be verified and delivered to very demanding performance requirements. Early in his career, his emphasis was on digital design where he has been awarded 4 patents in the electrical engineering area. As he moved into system engineering positions, he was engaged in all phases of programs from negotiating requirements with customer, developing system architectures, driving implementation, integration, and test, and finally product verification and sell off. His current position as system engineering manager affords him the opportunity to afford him to mentor engineers early in their careers on the importance of maintaining the vision for the delivery of the final product during all phases of the program.

David L. Melton, P.E. is an executive with 42 years of experience on military-aerospace, commercial and industrial programs in project/product development. He has presented several papers on the implementation of Taguchi Robust Design Methods and Quality Function Deployment in the System Development of products. Through April, 2005, he served as VP/Director of a space systems business. As of May 1, 2005, he formed an engineering consulting firm to support and assist small-medium size companies in building a business enterprise framework for the effective integration of new and complex technologies into their products/process. He specializes in the analysis, investigation and development of complex dynamic systems through implementation of system engineering practices including the application of scenario planning and simulation/modeling of system & business dynamics of complex systems applications. He is a Certified Taguchi Expert: American Supplier Institute (ASI) – 1999, trained in application of Quality Function Deployment: ASI – 1989, and a Certified Greenbelt: University of Michigan – 2003. He developed and taught three Graduate courses on Robust Product/Process Design.

Nancy Ridenour is a software manager with over 25 years of experience. Early in her career she developed real-time embedded software in the defense industry. Over the last 10 years she has managed the development of object-oriented software for both Windows and Real-Time Embedded applications in the aerospace industry. She also holds one patent and certificates in software architecture. She has a BSEE and MSE from Purdue University.

Martin Coe is a Systems Engineering professional with 29 years of product development experience. He has held numerous engineering/consulting roles in the development of more than 100 commercial products in the industrial instrumentation, consumer electronics and biomedical industries.

Martin has a B.S. in Electronic Engineering from Metropolitan State College of Denver and an M.S. in Systems Engineering from Southern Methodist University and holds certifications in Lean/ Design for Six Sigma and Certified Systems Engineering Professional.

His recognized work includes:

- Product development process improvements and product requirements for medical device organizations
- Several published papers / presentations on various Systems Engineering topics

Martin has worked with the IEEE, PDMA and INCOSE as an advocate for Systems Engineering in commercial product development and currently works as a Principle Systems Engineer in the medical device industry.

Made in the USA
Charleston, SC
11 February 2012